PRAISE FOR *LEADERSHIP FOR ENGINEERS*

"Bennett and Millam have built the bridge over the chasm between technology and leadership. With sound theory and illuminating practical experiences, they provide a roadmap for technology leaders to meet the 21st century challenges."

Arnold Weimerskirch
Honeywell Corporate Vice President of Quality (retired)
Former Chief Judge, Malcolm Baldrige National Quality Award

"This book is a must-read for practicing engineers and technical professionals to consider getting out of their comfort zone. Stop making excuses, explore yourself, dispel myths, and serve others by stepping up to leadership."

Tony Ramunno
Engineering Director
Great River Energy

"*Leadership for Engineers* is a timeless workbook for technical professionals who are interested in exploring methods for expanding their horizons and pursuing their passions. The book provides practical guidance on how to discover one's own true passion and then act on it for the betterment of one's business, for society, and for one's self."

Anne Coldwell
Engineering Director
Medtronic

"Great book and it fits an important audience that is often overlooked in leadership books. Many people want to lead, not everyone wants to be an executive."

Ellen Bösl
Senior Product Development Engineer
3M

"As a retired industrial scientist/engineer, I believe the authors have put together the keys to career development and career satisfaction for STEM professionals. Over the course of 47 years at Honeywell I learned many of the ideas, concepts, and tools that are described, but I have never seen them written down or in such depth. This book will meet an important need for individuals, industry, and the global community."

David Zook
Honeywell executive (retired)

"So often others see the potential in us that we don't see in ourselves. Finally, a decoder ring that helps us find our way! An easy read, with information that makes sense."

Dan Conroy
Vice President of Human Resources and Talent Management
Nexen Group, Inc.

"Great book to highlight how engineers, technical people, can have the presence and knowledge to lead from anywhere in the organization."

Jon Joriman
Senior Resin Chemist
Interplastic Corporation

"Leadership skills are increasingly important in both the hard and soft worlds of engineering. Engineers need to develop these skills to lead the increasingly complex and multi-disciplinary projects that are embedded in the Grand Challenges and to 'sell' these projects to owners and other stakeholders."

William E. Kelley
Director Public Affairs
ASEE (American Society for Engineering Education)

"*Leadership for Engineers: The Magic of Mindset* is a must read if you lead engineers, teach engineers, or are an engineer. Actually it applies to anyone with a technical background, not only engineers. This book is packed with insightful ideas and practical, real-world examples. The section on myths hits the nail on the head. And what I love about the book is that it not only identifies issues, it shows a path for individuals to find their own way to making a difference."

Louise M. Morman
Executive Director
Lockheed Martin Leadership Institute
School of Engineering & Applied Science
Miami University

"I have been an avid student of leadership for over 20 years, and wish your book would have been available to me in my early studies. While the book is targeted toward technical professionals, any young professional will greatly benefit from reading this book. It is a great overview on the topic of leadership, and loaded with many helpful exercises for those who want to develop leadership skills, and to truly understand what it means to be a leader."

Kathleen Kolbeck, PE. LEED AP
Retired President/CEO
Dunham Mechanical and Electrical Consulting Engineering

Leadership for Engineers
The Magic of Mindset

Ronald Bennett, PhD

Elaine Millam, EdD

McGraw Hill

Connect
Learn
Succeed™

LEADERSHIP FOR ENGINEERS: THE MAGIC OF MINDSET

1 2 3 4 5 6 7 8 9 0 DOC/DOC 1 0 9 8 7 6 5 4 3 2

ISBN 978-0-07-338593-8
MHID 0-07-338593-X

Senior Vice President, Products & Markets: *Kurt L. Strand*
Vice President, General Manager: *Marty Lange*
Vice President, Content Production & Technology Services: *Kimberly Meriwether David*
Editorial Director: *Michael Lange*
Publisher: *Raghothaman Srinivasan*
Executive Editor: *Bill Stenquist*
Marketing Manager: *Curt Reynolds*
Developmental Editor: *Katie Neubauer*
Senior Project Manager: *Lisa A. Bruflodt*
Cover Designer: *Studio Montage, St. Louis, MO*
Cover Credit: *© Wayne Calabrese*
Buyer: *Susan K. Culbertson*
Media Project Manager: *Prashanthi Nadipalli*
Compositor: *Laserwords Private Limited*
Typeface: *10.5/12 Times Roman*
Printer: *R.R. Donnelley*

Library of Congress Cataloging-in-Publication Data
Bennett, Ronald J.
  Leadership for engineers : the magic of mindset / Ronald Bennett, Elaine Millam.
    p. cm.
  Includes bibliographical references and index.
  ISBN 978-0-07-338593-8 (alk. paper)—ISBN 0-07-338593-X (alk. paper)
  1. Engineering—Vocational guidance. 2. Engineers. 3. Leadership. I. Millam, Elaine. II. Title.
  TA157.B46 2012
  658.4'09208862—dc23
                                                                    2012012043

www.mhhe.com

BRIEF CONTENTS

CONTENTS

DEDICATION

This book is dedicated to my wife, my best friend, Kathryn, for a lifetime of support, encouragement, and inspiration and to our three outstanding children, who lead by making the lives of those around them a joyous experience. The work would also not be possible without thousands of family, friends, and students who have enlightened me throughout my career; the alumni who gave freely of their time and experiences to demonstrate leadership at all levels; and two special mentors, Clint Larson and John Povolny, who throughout their professional careers and personal lives have inspired me and others to liberate their inner leaders.

R. J. BENNETT

I would like to acknowledge and dedicate this book to my husband, Don, who has been a lifelong supporter of my work in leadership development and teaching. I want to further acknowledge all of the many graduate students and business clients who have taught me so much about life, leadership, and how to plunge deeply into their own personal and professional pursuits, realizing dreams they originally felt were out of their reach.

E. R. MILLAM

FOREWORD

by George W. Buckley, chairman, president, and CEO, 3M

As a humble electrical engineer who over the years has been blessed with many opportunities to learn and to lead, I thank Ron Bennett and Elaine Millam for their very real contribution to an important cause: to help technical professionals discover and develop their own leadership skills. There is no doubt that early in my career I would have benefited from this practical, commonsense approach to leadership development.

I am convinced that the vast majority of people have more capability than they themselves realize. For some people, these talents emerge in times of trial and crisis; others simmer their skills over the years as their experience matures into wisdom. What all leaders have in common, however, is a belief in something better: a better approach, a better technology, a better enterprise, and even a better world.

To achieve the goal of something better, a leader needs to be comfortable with a level of risk taking amidst uncertainty. A leader needs to accept the responsibility of personally making decisions, even though some decisions will be unpopular in some circles. Other decisions may well define the success or failure of a project, or even a company. As daunting as that sounds, be assured that leaders are quite human. They learn through experiences—both good and bad—and they come to understand that confidence can breed even more confidence.

Along with confidence come courage and inspiration: the courage to do what's right in the face of uncertainty and criticism and the inspiration that causes opportunity and success to overwhelm the fear of failure. Successful leaders also learn that ego is an unaffordable luxury. It's better to have your shadow on results rather than your fingerprints. After all, leading is not the same as managing or micromanaging; the leader's mindset is not the same as the manager's mindset.

Leadership for Engineers: The Magic of Mindset speaks directly to this important distinction. I trust you will find it as enlightening as I do.

PREFACE

Most leadership books are written by and about CEOs. Many are inspirational and interesting, but they seldom provide tools to help others become leaders. However, we need leadership in industry, education, and government; we need leadership in setting public policy in our communities and country. We need leaders at all levels in every organization, not just in the executive suite.

In our view, leadership is the ability and courage to create a vision that inspires others, the ability to communicate that vision and to engage all the talent in the organization to focus on the same goal. This means you can lead no matter what your job title or position.

This book is based on our experiences in leadership training and education, as well as the experiences of many graduate students in the School of Engineering at the University of St. Thomas. These working adults entered the program as engineers, scientists, and other technical professionals. They shared goals of growing, learning, and self-improvement. Along with academic and technical instruction, they were invited to discover their potential, think broadly about their contributions to society, and create plans for developing their leadership capabilities.

INDUSTRY AND EDUCATION

During our careers in industry we have known many talented technical professionals. Although a few were satisfied with their personal and professional accomplishments, others clearly wished for more. Among these engineers, scientists, technicians, and mathematicians, many became disillusioned with their jobs. They felt disconnected from their organizations and did not see how their work added value.

They were not limited because they lacked technical capabilities, but because they had not developed leadership skills—or the courage and passion necessary to use them. In turn, they and their organizations were unable to benefit from these hidden or suppressed talents.

As we moved into academia, we have had the opportunity to create environments that help these professionals develop their leadership abilities, demonstrate their courage, and discover their passions. If this is possible in the classroom, we believed, it should be possible in business.

THE NEED FOR LEADERSHIP EDUCATION

For engineers, strong technical abilities can bring professional success—but so much more is possible. By developing leadership attributes, they can also realize increased recognition and personal satisfaction. More important, expanded skill sets let engineers make greater contributions to their organizations, their communities, and the world at large. This is reflected in the requirements from the Accreditation Board for Engineering and Technology (ABET).

The Engineering Accreditation Commission of ABET specifies criteria for all engineering programs. Criterion 3, Student Outcomes, requires programs to show that students attain 11 outcomes, often referred to as "a–k." This book addresses 6 of these outcomes, which are detailed in the appendix.

In their careers, many people educated as engineers move into management positions. Despite the ABET requirements for student outcomes, few are prepared to lead with confidence—and few are prepared by their companies to develop the skills and attitudes necessary to be good leaders in their organizations.

This book provides guidance on incorporating leadership into existing courses and activities, offering opportunities to assess and evaluate several Criterion 3 outcomes; it demonstrates the need to expand leadership education to practicing engineers, who are the emerging leaders in their organizations; and it provides suggestions for alternative approaches, serving as a resource for self-directed study.

With the generous assistance of the American Society for Engineering Education, ASEE, and the leadership of the Engineering Dean's Council, in 2009 we sent a leadership education survey Bennett and Millam 2012 to deans of engineering programs in the United States. All the respondents—100%—believe leadership education is important for engineers, yet only 46% include related courses in undergraduate programs, just 21% in graduate curricula. Those without leadership courses have asked how to incorporate the subject into programs that are already demanding. This book provides a place to start.

A DECADE OF DISCOVERY

Eleven years ago, as leaders in engineering education, we began a partnership with our constituents who were sending their engineering and technology professionals to graduate school. We spent time with them to learn what outcomes they valued most. Their answer was clear. They still wanted us to provide technical courses that introduced students to ideas and technologies that help them solve business challenges. But they also wanted us to help technical professionals see themselves as leaders and prepare to take on leadership responsibilities.

With that insight and tremendous industry support, we added three courses that spanned the graduate program. In this curriculum, students began with a base of self-awareness and a shared definition of effective leadership. They also conducted several self-assessments to better understand their potential. Through learning activities between each course, students developed plans for practicing leadership. Finally, they engaged in building team leadership skills, leading change, and broadening their understanding of global needs and challenges.

After nine years of experience with these graduates (Bennett and Millam 2011a,b,c; Millam and Bennett 2004, 2011), tracking and monitoring their progress, coaching and advising them along the way, we have observed phenomenal growth on their part. They have made serious choices and are challenging themselves after graduation to go further with their leadership practices. Each person's story is unique—and all have found a path forward to making a contribution.

These students come from many different backgrounds: business, medicine, law, information technology, engineering, and other technical backgrounds. They work in local, regional, national, and global organizations. They are technical professionals at the forefront of the global economic challenge, and they serve as a microcosm of technical professionals worldwide. This book shares their stories of exploration and discovery, and how they are meeting the challenges they experience in their own environments and in the broader world.

They have come to realize that leadership is not about their position or authority but, rather, how they serve and engage with others. They now think of an effective leader as someone who motivates others to reach shared goals. And they realize it takes people at all levels and with all capabilities to build strong and effective organizations. In this book, we share highlights of what they learned—and hope you find it equally inspiring.

Between us, we have more than 80 years of experience in industry and academia. Elaine Millam has worked with students at all levels, from K-12 to graduate school, and spent many years in industry as an executive responsible for leadership development and as a leadership coach. Ronald Bennett has extensive industry experience as an engineer, an engineering executive, a general manager, a business-to-business sales executive, an entrepreneur, and a postsecondary educator and administrator. Our experiences and viewpoints converge on one key issue: the need to maximize the talents and skills of technical professionals.

We wrote this book to help you achieve your professional and personal goals. It is meant primarily for technical professionals, technical managers, students of science and engineering, and others with education and experience in science, technology, engineering, and mathematics.

Every idea in this book was tested in a wide range of organizations and industries by people like you: people doing things that have led to extraordinary accomplishments and life changes. This book is about learning to think, feel, and act differently. It is about self-discovery, realizing that you already have most of what you need.

Not everyone wants to be a leader; we understand that. The concepts and experiences covered in this book will also help you recognize good leadership and participate in ways that make a difference. We assume you are motivated to contribute—to help make the world a better place.

INVITATION

We invite you on a journey with the technical professionals who shared their experiences and perspectives with us. Read their stories of growth, learning, and development. Look for their challenges, trials, and moments of achievement. Imagine yourself in their places; consider what you would do, and why.

Review the testimonies of emerging leaders, and let them inform your leadership choices and plans. You can use the scenarios in this book in a process of self-discovery, leading toward your own vision of effective leadership.

In the appendix, along with useful tools and exercises, you will find our contact information. If you have questions or comments, we want to hear from you.

INTRODUCTION

W hat goes into the making of your mindset? Have you ever stopped to consider what makes you think as you do? What does your inner voice tell you about yourself and your world? Do your beliefs keep you learning, seeking, and wanting to make a difference—or do they keep you from changing, growing, and letting go of the past? Whatever your history, you can still make a choice. Your mindset determines how you approach your life, and much of what happens in it.

As a technical professional in science, technology, engineering, or mathematics, you have unique abilities and opportunities to make a difference in this world. Your background and training are needed to create solutions to the monumental challenges of this century and beyond. Consider this list and ask yourself, "What is mine to do?"

- Make solar energy economical
- Advance health informatics
- Manage the nitrogen cycle
- Prevent nuclear terror
- Provide access to clean water
- Provide energy from fusion
- Engineer better medicines
- Secure cyberspace
- Restore and improve urban infrastructure
- Reverse engineer the brain

Beneath the emotional and political components of these issues are technical problems. As a technical professional, you have the knowledge, skills, and point of view needed to solve them. To make a difference, you need leadership skills and attitude. That means recognizing your innate creativity and combining it with the critical thinking skills that help you separate fact from fiction. We hope to capture your attention and spark your desire to make a difference.

Now is the time to invest in your future and position yourself as a leader. We want to help you develop a mindset that says, "I can and I will."

OVERVIEW

In this book, we share stories of emerging leaders and their experiences, as well as some critical information and tools that can help you, as a technical professional, reflect on your experience and decide what comes next.

You have untapped leadership abilities. This book will help you find and develop your inner leader, become the person you want to be, and pursue your passions in ways that are productive and rewarding for you and others. We identify some common myths to help you recognize false assumptions. Then we help you identify your own leadership skills, show how you can make a difference, and explain why the world needs your best professional self.

Each part of the book includes four chapters, with reflection questions at the end of each chapter. Take time to answer these questions for yourself. We promise that you will gain new insights into your thoughts, your mindset, and what is most important in your life.

Part 1—Exploding the Myths will help you confront some of the mistaken beliefs our society has about leadership and technical professionals. We show how others have challenged each of these myths and provide ways for you to let go of the mistaken beliefs that hold you back.

Part 2—Finding Your Inner Leader invites you to begin a self-assessment, discerning your strengths, talents, possibilities, and potential. It encourages you to analyze your wants, needs, and desires—and then create a plan for developing yourself as a leader.

Part 3—Making a Difference asks you to assess your professional responsibility and your obligation as a technical professional. It invites you to "be the change you wish to see in the world" and provides tools and exercises to help you become just that.

Part 4—Why the World Needs You speaks again about the global demand for innovation, critical thinking, and systems thinking and functioning. It restates the challenges of this century and beyond, showing why combined technical and leadership skills are so important.

PART 1

Exploding the Myths

This section uncovers and explodes 20 familiar myths that operate within our minds, our personal lives, our organizations, and our society.

In this book a myth is not a parable from Mount Olympus; it's a commonly held misconception that restricts our imaginations and ambitions. Some are originally based in fact, although they have become less accurate over time. Others have never been true but evolved from unrelated stories within our culture. They still affect workplaces and careers today.

As you read through these myths, think about which ones seem like beliefs. Follow the trail into your own past and you can discover how those beliefs—those myths—have affected your thoughts and actions. You may realize where they began, and whether they are still relevant.

The first five myths keep us thinking that, as ordinary people, we cannot possibly lead others or make a significant difference. Other chapters will help you explore myths about yourself, leadership, families, organizations, and society. All myths start somewhere; some will stop here. ●

Myths about Ourselves as Leaders

Myth 1: One Person Cannot Make a Difference

This is a peculiar belief, since it has been disproven again and again. Consider the impact each of these people has had on our understanding of the world and how we live:

Alexander Graham Bell	Michael Faraday	Antoine Laurent Lavoisier
Niels Bohr	Enrico Fermi	Guglielmo Marconi
Nicolaus Copernicus	Henry Ford	Gregor Mendel
Seymour Cray	Benjamin Franklin	Isaac Newton
Francis Crick	Galileo Galilei	J. Robert Oppenheimer
Marie Curie	Robert Goddard	Louis Pasteur
Charles Darwin	Jane Goodall	Linus Pauling
Rene Descartes	Stephen Hawking	Max Planck
Rudolf Diesel	Grace Murray Hopper	Jonas Salk
Bonnie Dunbar	Johannes Kepler	Nikola Tesla
Thomas Edison	Jack Kilby	Leonardo Da Vinci
Albert Einstein	Alfred Kinsey	Eli Whitney

Not only did these people discover or create things that changed history, they all had technical training in science and engineering. You may not recognize every name on the list, but most of them should be familiar. Add one more scientist to the list—anthropologist Margaret Mead, who is credited with a keen observation about human nature and innovation:

Never doubt that a small group of thoughtful, committed citizens can change the world. Indeed, it is the only thing that ever has.

Not everyone will change the world—but you can definitely make a difference. Among the people we interviewed, with experiences and backgrounds that are common to many technical professionals, several took the personal initiative to create change when things weren't working.

Carol Jacobs found that her colleagues in a major industrial and consumer products company were reluctant to speak up or ask questions. Carol spoke up and got results; others now see her as a model. Wade Dennison identified a major structural problem in the medical device industry, then organized a statewide partnership among industry, academia, and government to resolve it. Bea Ellison made radical changes in the product development process at a top technology company, reducing cycle times and increasing organizational involvement. Here is her story:

Bea Ellison, an emerging leader at a major industrial company, was asked to meet with customers to gain insights and inputs about future needs. Having had little customer contact as a project manager, she was excited by the opportunity. The customers asked for a product that required a whole set of new technologies, and she was assigned a lead role for the project. Within six months of the project launch, she found her team expanding from 2 to 60 members.

Because of the size of the team, the communication challenges, and the many sites involved, Bea held a two-day kickoff meeting at company headquarters. There, all the team members were briefed on the project background, learned from each other's expertise, and developed a road map for the product's introduction. They established performance criteria and defined roles and responsibilities. Team members gave fantastic feedback about the process. They felt fully engaged with other team members and were thrilled to start from a common understanding of the challenges ahead.

As this story shows, one courageous person can initiate changes that affect a whole organization. Inviting others to join in a process can have significant impact on far-reaching directions. Bea's actions can inspire others to notice opportunities and take action. A mindset can be changed.

Myth 2: "I'm Just a Technical Person, Not Qualified to Be a Leader"

Many use the excuse "I'm just an engineer or a scientist and don't have the credentials to make a difference" for not doing what they know is needed. Yet they are on the front lines—the best place to identify problems, and to lead in finding solutions. It is precisely the people who know the technical facts and are close to the problems that can make a difference.

Technical professionals have often driven change, from Seymour Cray challenging conventional computer design to Roger Boisjoly, the engineer who tried to prevent the launch of the space shuttle *Challenger* in 1986 because he correctly believed critical seals would fail. As a technical professional, you have personal power, which is greater than position power. You have influence in the informal organization that gives you leverage.

Imagine applying your technical skills to designing earthquake-resistant structures or more effective wind energy devices. Picture a genetic engineer manipulating organisms to help solve world hunger or a team of biologists and engineers who create life-saving medical devices. Mechanical engineers design and build machines that address human needs—and wants. Remember that the people who invent consumer electronics, web applications, and other conveniences are engineers as well. They all make a difference.

Here's a story that shows how important personal power can be:

> Dan Jansen, a program manager at a defense manufacturer, had no direct reports for most of his career, yet people throughout his organization saw him as an influential leader. He ran virtual organizations with just personal power. This, coupled with attributes that demonstrate respect, a history of treating people fairly, impeccable ethics, and responsiveness to his customers, helped him build the trust that became his platform for leadership. Dan has said that if young people seek to learn philosophy and ethics, and learn how to be true to themselves, they will recognize their personal power more easily and be better able to use it to build healthy, strong relationships.

Dan refused to believe he was "just an engineer." He saw how he could exercise influence, acknowledge his power, and help others find their power as well. He provided leadership from a position not of credentials but of credibility.

Myth 3: Scientists and Engineers Lack Training to Be Leaders

Few people associate technical expertise with leadership, mostly because of the training that prepares young adults for careers. Not many college students with passions for science and engineering are also attracted to business administration, which provides a common path to management and executive leadership. Likewise, students with hearts and minds for business seldom pursue technical subjects at the same time.

But consider a broader description of leadership and the connection makes better sense. A leader can be described as "someone who will take you to a place you wouldn't go by yourself." A scientist seeks to discover—to go where others have not. And an engineer works to create things that have never existed before. Those things sound a lot like leadership.

Then consider the paradox within business management. One of the most popular books for executives is *The Prince* by Italian Renaissance author Niccolò Machiavelli. He counseled readers to practice image management and political flexibility, and to avoid change, if possible: "there is nothing more difficult to take in hand, more perilous to conduct, or more uncertain in its success, than to take the lead in the introduction of a new order of things."

Yet in business, there is nothing more dangerous than *not* being willing to change. Besides, technical professions are constantly changing our lives: what we know, how we improve the human condition, and how we treat the earth's limited resources. The role of the technical professional is to create change. Here's an example of that principle in action:

> Rae Collins, a product development manager in managing printing and imaging products, was willing to create change. She found positive challenges in a male-dominated industry, working with new technologies, constant change, and high stakes. She managed suppliers differently than she was taught, but in a way that built strong relationships. Her team was developing products that had never existed before, but their competitors were moving quickly as well. She learned to manage upward in her company, take risks without fear, and be true to herself. This leadership helped her team prevail—and earn recognition for it.

Effective leaders, like Rae, need to believe in themselves, enlist others in leading change, and do what they know is right. Here's another illustration of personal change leading to professional growth, in unexpected ways:

> As a capable R&D staff member in a manufacturing firm, Jerry Johnson was given an opportunity to lead a major business unit. Shortly after his promotion, however, the business experienced a major downturn and Jerry was laid off. He was shocked to see his rising star suddenly fall. He moved quickly to take charge of the situation and formed an LLC organization with friends from his former company, working to develop an innovative product for the military. Beyond this, he volunteered to help people in Haiti and South America. He placed hand- or foot-powered grinders in communities to help people start small businesses and improve their life conditions.
>
> Jerry is an entrepreneur at heart. He built confidence and tested himself through these experiences, and he built a large network of people who have been helpful in many ways. He has since been invited back to his original company to take a leading role in engineering and has decided not to do that—at least not yet. Although he loves fixing problems, he is now dedicated to serving others, making the world a better place. He feels that, to lead, you have to keep taking on new challenges and learning new things. That describes exactly what he has done.

Myth 4: If You Challenge the Status Quo, You Could Be Fired

You have a family to support and mortgage to pay. True. You might have trouble getting hired somewhere else if you get fired from where you are, so it's best not to rock the boat. False. According to a venture capital executive we met, if being fired were the end of a career, there wouldn't be a single CEO in the electronics industry. The real issue is not whether you contradict the current practices in your organization but whether you create value by doing so. It helps when your supervisor is supportive:

> As a quality engineer in a major medical device firm, Corrine Anderson found her first and most meaningful leadership role after attending training on just-in-time (JIT), a new concept at the time. She approached her boss and suggested forming a grassroots effort to implement the practice. He agreed. She formed a team of interested colleagues and explained her goals. They executed the plan, bringing about a 50% reduction in cycle time and large inventory reduction, which helped the team earn a company award. This built Corrine's confidence, which has continued to serve her—and her company—ever since.

Corrine was not afraid of challenging the status quo. She took action, seeking support and engaging interested colleagues. Hank Bolles had a similar experience making waves:

> Hank found his company had moved to a "command and control" model that did not suit him. A former boss had moved to another company, which was beginning to adopt cellular manufacturing methods. This boss recommended Hank in the new organization; because of his experience and success, he was hired as a project manager. He applied his knowledge from his former experience to robotics, lasers, and automated material handling. He became highly visible to upper management because he was passionate about bringing about change.
>
> At the same time, he was completing a master's degree in manufacturing systems. He let top management know he wanted to do something more strategic. The company was looking for help in competitive intelligence, business research, and acquisition analysis and wanted someone with analytical skills and an understanding of operations. They chose Hank. It was

(continued on next page)

(continued from page 7)
an entirely new role, combining financial planning and analysis with peer analysis between companies. This was a new function for the company; no one could define it exactly. With help from a mentor, Hank learned to put things in terms that mattered to the company's executives. He brings value by condensing volumes of complex data to create useful insights. His technical background helps him understand detailed information, and his leadership experience helps him turn it into wisdom.

Both Hank and Corrine were passionate, and they knew what their companies needed and how to engage with people to get their support, demonstrating how to add real value. Still, the fear of being fired keeps many technical professionals from speaking up, even when they see a problem and have ideas about how to solve it.

The fact is, more people are fired for taking too little action, failing to contribute, than for trying to improve the way their organizations function. Here's another person we interviewed who stood up and challenged conventional thinking, only to be rewarded with additional authority and responsibility for showing initiative.

Dan Jansen had a personal and professional dilemma. He had material review board responsibility to accept or reject incoming material. One lot of a key component had failed. If he rejected it, his company would fall further behind on a major Defense Department project that was already late. Despite management pressure to accept it, and a concern that his decision might cost him his job, Dan did what he believed was right and rejected the material.

In a conference call with the defense contractor customer and the Air Force, a notoriously tough negotiator with the contractor demanded that Dan explain his reason. He did. As Dan was told later, the negotiator jumped on the table and yelled, "AT LEAST SOMEONE HERE KNOWS WHAT'S GOING ON!"

He thanked Dan and had him leave the call. Then he and the Air Force proceeded to berate the managers. Dan not only kept his job but has continuously been given more responsibility and now leads the company's initiatives to improve quality and delivery with all its major commercial airline customers worldwide.

Forward-looking organizations want people who create change. But habits are difficult to break, so even changes top management claims to want can be difficult to bring about. So, if a company or an institution fires a technical professional for doing the right thing, the problem is with the organization. This experience, however it turns out, helps you develop confidence and courage—and with those characteristics, you can do anything.

Myth 5: The Riskiest Thing You Can Do Is Take a Risk

Although change may be risky, lack of change is riskier. Successful organizations realize that change is coming at them every day. Leaders within those organizations need to know what is happening internally and externally, and be ready to react quickly and with confidence. Their organizations need to be nimble and prepared for rapid changes. Those who cannot or will not respond to changing conditions will not last. We have all seen such companies left behind.

One definition of risk is the quantifiable likelihood of loss or lower than expected returns. Notice that a decision not to change has its own risks; an entire industry can be transformed or rendered obsolete by a new technology, or by a shift in consumer habits. One that keeps generating the same products and services may lose market share even without a direct competitor taking it away. Here's an illustration:

Take the top 50 companies from the Fortune 500 list in 1955. These were the most profitable, productive enterprises in the country—and this wasn't very long ago. Between then and now, the United States has had its two longest periods of economic growth. You might expect most of them, if not all, still to be successful today. But if you did, you would be wrong.

In 2010, only 7 of those top 50 companies were still on the list. Of the top 10, only 2 remained. That means 82% of the strongest companies sank in 55 years, and just 18% stayed afloat. Books such as *Good to Great* by Jim Collins and *The Innovator's Dilemma* by Clayton Christensen provide clear examples of companies that have persisted and others that have failed. Many that remain have changed drastically to stay aligned with changing demands.

It's reasonable to assume that more of these companies failed because they did *not* take risks than because they *did*. But the real lesson here applies to the future. Success today does not even guarantee survival tomorrow.

Almost any action is better than none. It is even more likely to succeed if it is thought through, is based on real needs, and has clear organizational value in the long run.

Chapter 1 Reflection Questions

1. When and where have you taken risks in your life?
2. What happened when you took a risk, and what did you learn?
3. Where are the opportunities for you to speak up in your organization?

4. How can you create changes that help your organization better meet its goals?
5. What does it mean for you to find and exercise your personal power?
6. When you exercise your personal power, what are the results for you and your organization?
7. How do you create value for your company? How do you know?
8. What myths have their grip on you? Where did they come from?

CHAPTER 2

Myths about Leadership

Myth 6: Leaders Are Born, Not Made; Some People Just Have It

> Becoming a leader is synonymous with becoming one's real self. It is precisely that simple and also that difficult.
>
> —Warren Bennis, leadership studies pioneer

This statement became the platform for developing a series of college-level leadership courses. It builds on the notion that, to develop yourself as a leader, you must examine your beliefs, talents, skills, interests, motivations, and visions of what you want to create in your life. It completely contradicts the belief that some people just "have it." Leadership is developed, practiced, and nurtured through daily experience and support from others.

In our interviews, we found that emerging leaders were given more responsibility because of their demonstrated ability to get things done. Some were never told why they were chosen. Few were given tools or guidance by their organizations to make them better leaders. Those who were became more consciously competent and knew why they were chosen, and they were strong initiators of their leadership moves. They influenced their organizations to make the right decisions.

Betty Jarrett, an emerging leader at a high-technology company, has had several positions over her 22-year career, ranging from supply chain specialist to IT leader, with many variations of all of these roles. She is a black belt Six Sigma leader and has been put in charge of many challenging projects, moving from one project to the next every 18 months or so.

Thanks to her high energy, focus, and imagination, she has had many successes at her company and has received many awards. She is goal-oriented and confident that choosing the most responsible course of action now will bring benefits later.

Betty believes in developing others to their highest potential. She is friendly, assertive, perceptive, and candid—as well as quick to give credit to her team. As a high achiever, she takes responsibility for her own development as well. These are not genetic characteristics but learned skills, just like overcoming fear.

We learn in many ways—some by trial and error and others more deliberately from the lessons of others. We also learn through a combination of gaining self-awareness, learning from others, trying newly learned skills in real time, getting feedback from a support group, and continually striving to improve our abilities, build strong and healthy relationships, articulate personal and organizational visions, and inspire others to share in achieving that vision.

When these alumni we interviewed first saw themselves as leaders, the experience was frightening, audacious, and inspiring all at once. They had never thought of themselves as leaders, never truly noticed their capabilities, and wondered aloud why they had believed so many of the myths. As they put these skills into practice, however, they discovered that they could rise to any occasion, noticed their desire to make a difference, and found their own unique ways of making things happen.

> Be not afraid of greatness: some are born great, some achieve greatness, and some have greatness thrust upon them.
>
> —Malvolio, from Shakespeare's *Twelfth Night*

Myth 7: You Need a Title to Be Called a Leader

Within most organizations, a title indicates both your functional role and your position relative to others; no wonder we associate leadership with rank. But we have seen effective leadership in every part of an organization. You probably have, too, so let's put this myth to rest.

There is no one right way to lead others, nor is there a prescription for how to lead. It is a process of learning and developing yourself. Start by knowing clearly who you are and recognizing your gifts and talents, skills and knowledge, interests and passions, areas of strength and shortcomings. Then determine what influence you want to have in your environment—whether that is your home and family, your social environment, your workplace, your community, or the world at large. You can be a leader from wherever you stand by communicating clearly and honestly, helping others become powerful, recognizing how to make things happen, and working with integrity.

> "Know thyself" was the inscription over the Oracle at Delphi. And it is still the most difficult task any of us faces. But until you truly know yourself, strengths and weaknesses, know what you want to do and why you want to do it, you cannot succeed in any but the most superficial sense of the word. The leader never lies to himself, especially about himself, knows his flaws as well as his assets, and deals with them directly. You are your own raw material. When you know what you consist of and what you want to make of it, then you can invent yourself.
>
> —Warren Bennis, *On Becoming a Leader* (1989)

A title doesn't make a leader and never has. Lech Walesa is a good example. As an electrician at the shipyards in Gdansk, Poland, he spoke up on behalf of his fellow workers, neighbors, and citizens. His passion for greater economic and political freedom made him a leader in changing his country's government. Eventually, he became its president—but he was a leader long before he had any title at all.

When developing leaders were asked to identify the people *they* considered leaders, most of them named a relative, a teacher, or someone else from their own past who had no title. These people were influential because they provided confidence, support, and encouragement. By example, these mentors showed how to live with integrity, help others in caring ways, and seek accomplishment whether or not it would bring recognition or acknowledgment.

In our interviews, we found many examples of leaders rising from within the ranks of their organizations. Here are two of those stories:

> Bobby Bridges was an engineering manager at a truck assembly plant. For many years, the corporate quality group had tried to establish a top-down process to monitor and correct cab welding problems, but it never caught on. Bobby had developed an improved process to do this in his plant. Now every weld could be traced to a specific machine and tool, so defects could be detected and corrected quickly. He shared this with his colleagues in other plants, and word spread. It was readily accepted and is now the corporate standard. Bobby simply shared the methods he developed and was recognized companywide as the leader of this initiative.

> Nate Keyes was hired as the vice president of manufacturing at a company that manufactures high-end tooling. As good a company as this was, there was work to be done, and his personal leadership skills would be tested. He recalls his first meeting with the manufacturing manager and asking, "How's your quality?" The manager answered, "It's so good we don't measure it."
>
> Nate suggested to the manager that he get some bright orange buckets and place them around the plant. If by chance there should be a defect, the part could be put in the bucket. When the bucket became full, they would place it in the front entrance for all employees to see. It didn't take long to fill one bucket, then two, then many. One of the seasoned manufacturing people stopped by Nate's office and said, "I think you're on to something." He helped the employees discover the problem for themselves and created an environment for them to solve it. Although Nate had position power, it was his personal power that made the difference.

Leadership is a process of influencing the activities of a group to achieve mutual goals. It happens in many ways, some less inspiring than others. This leadership definition can be driven by an authoritarian leader, a leader who believes in shared power, a quiet and unassuming leader, or any other type.

Sometimes leadership is not developed in the work environment but through volunteer experiences. Dick Bastion was an engineer at a mainframe computer manufacturing company. His boss was very supportive and had taught him to appreciate what everyone contributes. He wanted to honor her in some way—and his chance came soon.

Dick had joined the Jaycees, where he got a different perspective on leadership. His first project with the Jaycees was to participate in a "Boss of the Year" award ceremony for bosses who mentor or teach their employees. He nominated his boss and asked how he could help. To his surprise, he was put in charge. He had support from members of the board, and the event was successful.

Although this forced Dick outside his comfort zone, it was a positive experience. Since then, Dick has taken on advanced leadership roles in the organization, up to the state level, and has developed speaking skills that have served him well in both volunteer and career advancement. He has realized that his primary motivator is giving back to society. Through his many volunteer activities, he gets the great satisfaction of knowing that he makes a difference.

Think about what kind of leader this definition suggests, and notice what appears to be effective in these stories.

Myth 8: Leaders Tell Others What to Do

The best leaders are active listeners who lead in a Socratic style by asking questions and further developing their inquiry skills. They know their colleagues have more and broader knowledge collectively than they could possibly have as individuals, and they create the environment and conditions for all to perform as a team.

A classic example of this collective leadership is reflected in the Value Creation Model. It demonstrates how organizations create value for multiple stakeholder groups. Motivated employees create innovative processes that result in exciting products, which generate repeat business, while developing lean operations that cut costs. The results are delighted stakeholders and an organizational impact on society. This can happen on a consistent basis, however, only if the organization knows how to create and sustain a culture that encourages innovation and supports employees in trying new things and taking risks.

This model was created at Honeywell, developed with engagement from many leaders and followers in the organization. They collaborated to show how all stakeholders must work together in a synchronistic fashion—not necessarily having leaders tell others what to do.

The Value Creation Model offers an opportunity for everyone in an organization to better understand their interdependence, as well as their potential impact on society. It represents the possibility in an organization when all are in synchronicity with goals and voices. (See Figure 2.1.)

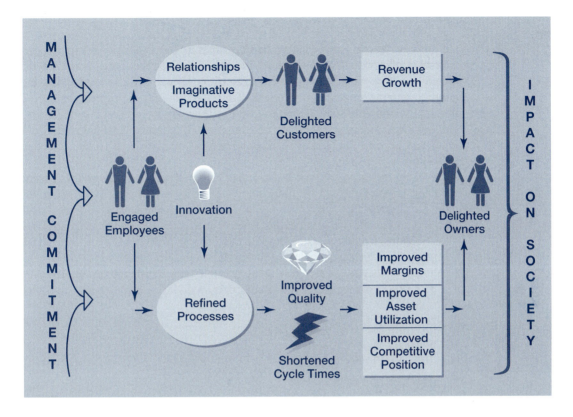

FIGURE 2.1 Value creation model.

In previous accounts, we mentioned the approach of many emerging leaders. They talk about the importance of listening and creating mutual respect. They notice themselves becoming annoyed with others who are "upwardly focused," playing politics. Ellie's story reflects how she exercised an active listening approach, rather than telling others what to do:

> As the leader of a small team in a medical device company, Ellie Fitzgerald was confronted with a situation in which four team members each had different points of view on how to handle a specific situation. Ellie set up a two-hour meeting and stood at the whiteboard, laying out the pros and cons of each approach by asking the team questions and documenting their responses. Doing this systematically helped the team members recognize the appropriate path, and all she did was facilitate the discussion in a productive manner. The team agreed on one approach and, as they left the room, all felt their ideas were heard and respected.

Myth 9: Only Extroverts Can Be Good Leaders

Because engineering and science attract many introverted people, this myth contributes to the mistaken impression that they cannot be leaders. For a dramatic example in both public and private life, consider former president Jimmy Carter, a nuclear engineer and an introvert. He brings together people from all countries and socio-economic status levels, as well as those divided on almost any issue, helping them understand each other's goals and collaborate to achieve them. All types can become leaders, as there is no prescribed style or type that is better than another.

Among the technical professionals and students we have known, many have used the Myers-Briggs Type Indicator (MBTI) to gain insight into their own thoughts and feelings. This personality assessment questionnaire is designed to measure preferences in how people perceive the world and make decisions. The classifications, called types, are based on theories of psychologist Carl Jung. The assessment tool was created during World War II to help match women who were new to the industrial workforce with the wartime jobs for which they applied.

The MBTI personality assessment exercise divides people into two groups for each of four characteristics:

- Extraversion (E) or Introversion (I)
- Sensing (S) or Intuition (N)
- Thinking (T) or Feeling (F)
- Judgment (J) or Perception (P)

These classifications are not absolute; they represent scales from one to the other, for a total of 16 potential types. We found that young adults in engineering and technology fields were spread across all 16 types, but with a few definite trends. There were more introverts than extroverts; these introverts are the thinkers, who prefer to take their time before they speak out. They tend to be energized by time alone for reflection, to work through their thoughts before acting on them. They like concentration and reflection, focusing on the inner world of ideas and what could be. Many of these introverted characteristics actually help make people effective leaders.

Other useful assessments are learning styles, social styles, and emotional competencies. In another section, we discuss Driver, Expressive, Amiable, and Analytical social styles and learning styles that are often related to MBTI types. Different assessment tools can bring similar results. For example, extroverts are essentially expressive, with characteristics of high assertiveness and high responsiveness. Amiables are highly responsive but low in assertiveness. Dwight D. Eisenhower was an amiable—but he was chosen to lead the Allied forces in Europe during World War II, instead of the extroverted Patton or Montgomery, because he knew how to keep focused on a goal and to provide leadership for the strong personalities who would execute the war plans. Eisenhower had the required set of characteristics for the situation.

Orrin Matthews is an introverted leader who holds people close to his heart. With careful listening to his team members as standard procedure, Orrin learns what makes his people tick. He wants to see them succeed, to stand out in his company; he has earned a solid reputation for fairness and generosity with those at all levels in his company. He is also modest in explaining how he gets the best from his people, as well as how he influences the corporate culture of his organization.

He also recognizes a possible downside to his approach. Because he wants to give people the benefit of the doubt, he may overlook or enable less productive behaviors. However, he sets an example for others, showing how much he values continuous learning. He expects them to keep reading, trying new things, bringing forth ideas for new projects, and avoiding complacency. In return, the people who work with him consider him an inspiration and give him their full respect.

Countless others with a wide range of skills, abilities, and preferences have become leaders, if less well known. They come from all social styles and preferences—but they understand what motivates people and how to engage others in creating a shared vision. These leadership characteristics are well demonstrated in the people we interviewed.

Myth 10: Leadership Means Position Authority

Many, if not most, technical professionals enjoy direct contact with their areas of expertise. Scientists like to be in laboratories, for example, and engineers at test benches. By contrast, people with management titles and supervisory authority are not expected to have their hands on technical subjects. It's generally not required for executive positions. But this contributes to the misconception that people with leadership titles and those with technical knowledge have completely separate sets of skills.

This myth is a close cousin to the myth about title, with the addition of authority—a belief that someone needs to act as a figurehead. But positional authority does not guarantee leadership. In fact, because positional authority is usually dominated by the need to manage budgets, projects, and logistics, it reduces the time and energy available for leading people, being creative, engaging others, and aligning teams. Management and leadership are complementary skills—but not all managers are leaders.

Rather than demanding or commanding, leaders engage with others in their organizations to achieve mutual goals. Leadership is about collaboration, teamwork, and innovation that sparks new ideas, new technologies, new inventions, and even new ways of working together. It thrives on cross-functional teams, respectful conversation balanced with constructive conflict, personal integrity, and challenge. Your

Mac Casey, a quality engineer at a major industrial organization known for innovation and global leadership, has experienced some tough times in the past three years. He has survived four rounds of layoffs and has found himself pushed to meet increasingly difficult requirements: cutting costs, responding to demanding OEMs in his industry, and picking up the load of 40% staff reductions.

This has been a learning time for him in many ways—how to adapt and remain flexible, to take on extra assignments even as the workforce is demoralized. He is thankful for graduate courses he took, learning how to lead by example and through influence. These two key skills recently saved the day for him.

Regardless of the challenges in his workplace, he conducts business by using influence management, facilitating problem solving and conflict resolution, and leading a team of internal people. He has no position authority over his suppliers or his project teams but leads through influence with these varied groups.

His leading plan was to make a difference in people's lives and to become more skilled in helping others set priorities. He works to help people stay motivated and challenged, and to make significant contributions, while realizing their dreams. He is still learning as an emerging leader—and is grateful he can prove himself more fully through his experiences.

"boss" may have more power and more say than you do, but your boss and your organization need you, your ideas, and your full engagement in order to succeed.

Many of the developing leaders we interviewed talked about the way they had transformed their own thinking and, therefore, their actions as true leaders in their organizations.

These examples show an effective approach for leadership. Instead of being the authority who says, tells, or sells, this kind of leader gains buy-in and gets people involved. It takes a special set of skills to invest in people, learn what makes them tick, and lead them to discover their greater capabilities. In organizations with leaders like this, the results are easy to see. They engage others, encourage them to consider new possibilities, and bring their full selves to work.

Myth 11: Technical People Work with Things; Leaders Focus on People

Although technical people *do* work on things, and *did* work alone in the past, that is no longer the case. Today they work in teams, which are based on people and relationships. They need to engage others through oral presentations, project reports, and technical analyses, all of which require communication skills. Some technical people are already good at them, whereas others need to work on their skills—but all technical careers require people skills.

Dr. Joe Ling, former head of the Pollution Prevention Pays program at 3M and member of the National Academy of Engineering, had a saying: "Environmental *issues* are emotional; environmental *decisions* are political; environmental *solutions* are technical." Although the answers to environmental and other questions may be technical, which is the domain of technical professionals, they must also know the emotional and political aspects of the issues and be able to deal with them. Here's a technical professional who learned the importance of people skills:

When Dan Jansen was a boy, he would ask his father what he would become when he grew up. His father's answer was always the same: you will be a salesman. Dan hated this response; he thought of auto salesmen, encyclopedia salesmen, and other models of salesman at the time and he didn't want to do that, so he stopped asking the question.

Years later, as an engineer convincing management to spend $175K on a project, he realized that he was a salesman. He recognized a need, understood the value of meeting that need, and believed it was good for the company. He had discovered the four basics of the selling or influencing process—trust, need, help, and decision anxiety. His supervisor in the 1990s taught Dan how to make a presentation to management with this advice: know your audience.

Now Dan is an engineer and program manager. When asked what he does, he responds, "I am a salesman." In the initiatives he champions, he has to sell his ideas. His skill as a salesman enables him to direct the power of his technical skills toward solutions by engaging others on his team.

Once they acquire leadership skills through work experience and graduate courses, many technical professionals become inspired to pursue doctoral degrees or career advancements. Often this choice reflects a desire to influence others and to play a broader role in society through both leading and learning.

Cal Archer has been a practicing engineering leader for 26 years and is proud of that fact. He moved to Minnesota 10 years ago and works as an engineering manager at a major defense manufacturing company. He has held many leadership positions with the Institute of Electrical and Electronic Engineers (IEEE), the Minnesota Society of Professional Engineers, and the American Institute of Aeronautics and Astronautics and plans to continue those relationships.

(continued on next page)

(continued from page 19)

However, he is clear about his need to "give back" in ways that he sees as useful to society. Hoping to guide a younger generation to lead in new and different ways, he has begun a doctoral degree program in organizational leadership.

He wants to keep his mind open to new insights, and he finds that reading and writing help him pursue wisdom and avoid cynicism. He is full of energy and hope for the future. His goals for teaching, mentoring, and serving in professional organizations are critically important to him.

Cal is quiet and unassuming, but his gifts are readily apparent; he knows how to reach others and bring out their best. He is a team player who makes sure others get credit for their contributions. While focused on his own learning, he stays aware of news in his organization, his industry, and the world at large.

Chapter 2 Reflection Questions

1. Whom would you name as the most admired leader in your experience?
2. What characteristics made this person influential in your life?
3. How did this person make you feel?
4. How do you define effective leadership or its ingredients?
5. Think of an experience you had with a powerful leader. What did you learn?
6. How have you tried to use that knowledge in your own leadership practices?
7. What leadership practices keep you thinking and growing?
8. Who are the supporters that keep you on track and hold you accountable?
9. If you developed your own leadership model, what would be the most critical variables?

CHAPTER 3

Organizational Influences

Myth 12: Organizations Won't Treat Technical Professionals as Leaders

This myth is tied to several others, such as the one about technical professionals preferring to work alone or the one that says they lack interpersonal communication skills. These myths all come together in the workplace, where they become part of an organization's culture. As a result, few technical professionals are considered well suited for leadership, even in their own eyes. Facts contradict this and other myths. Actually, many engineers and scientists are promoted to leadership roles (National Science Foundation 2003) (see Figure 3.1). Once they hold these positions, they are known as managers, although they seldom receive people management training. What they learn about management often comes from the role models around them—for better and worse. Some are lucky enough to have mentors, but most learn through trial and error, seldom seeking help from others.

When engineering and science leaders succeed, they may not attribute their success to their initial training background or career choices. For example, in Bill George's book *True North,* he neglects to mention that he is an engineer by training. In similar books, leaders with science or engineering backgrounds are referred to as managers, founders, or entrepreneurs; their technical history is invisible. One exception is Seymour Cray of Cray Research, who kept the title of engineer along with founder, entrepreneur, CEO, and chairman.

Statistically many science and engineering leaders became managers and leaders in their organizations. As you consider your future and your organization, think about this: the National Science Foundation reported in 2010 that more than 20% of graduates with degrees in technical subjects move into management roles within 30 years. (See Figure 3.1.)

Of course, there is a difference between managing things and leading people. We found that most technical professionals believed their organizations were focused on things, including profitability and shareholder value. As we discussed the differences between management and leading, technical professionals told us that serving as a manager was expected and rewarded. This meant it was important to stay within budgets, focus on short-term outcomes, accept the status quo, pay attention to systems and structure, and be a good administrator.

When we discussed leadership, however, they emphasized originality and innovation, focusing on people and processes, having a long-range perspective,

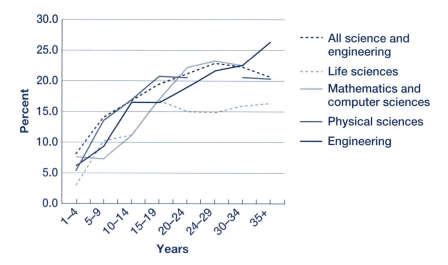

FIGURE 3.1 Science and engineering bachelor's degree holders in management jobs since degrees, 2003.
Source: National Science Foundation, 2003.

asking what and why rather than how and when, challenging the status quo, developing others, inspiring trust, and doing the right thing rather than doing things right. Many authors have written about these distinctions (Bennis and Nanus 2003; Kotter and Rothgeber 2006; Rosen and Berger 2002), yet few organizations operate as if success required both—as if effective outcomes depended on true leadership. Some of the technical professionals we interviewed are showing the way. You may be one of them; here's another:

> Sam Allen, a research and development manager with a medical device company, felt he made clear distinctions between management and leadership. After studying both concepts in theory and practice, he says, he now understands how leadership can and should be exercised at all levels in the organization.
>
> Focusing on continued self-awareness was a key concept for him. He has used that understanding to craft his leadership strategy, and he uses clear leadership principles and values in his personal and professional life. Leading by example is extremely important to him; he firmly believes in walking the talk: "I would not ask someone in my organization to do something that I would not do myself."
>
> Sam is proud of helping the R&D group become a patient-focused organization. He worked to get his engineers into hospitals, see their products in use, and get ideas and comments from end users. He has worked to improve access to training and secure company funding for it. He is proud of his group and feels that by combining ethics, leadership, and passion he has transformed his organization.

Effective leaders create productive organizations. By articulating their vision, mission, strategies, and tactics, they set the stage for making the right things happen.

Myth 13: Organizations Isolate Engineers from Leadership

People in engineering and science are often seen as belonging to a separate class from those in administration and management. In most cases, however, this happens only if technical professionals let it happen. To counter this myth, people in technical fields need to think as entrepreneurs—because they are. As an entrepreneur, you are the sole proprietor of your career and in charge of your life. You choose what you want to do and how you want to do it. No one speaks for you except yourself. Here are the things you must do to assume control of your professional destiny:

- **Assert yourself.** Speak up, speak out, and let others in the organization know about your interests and abilities.
- **Adopt systems thinking.** Ask how you can add value to your organization, how your job relates to the enterprise, and how your organization fits into the global economy.
- **Create a team.** Develop relationship-building skills to share your ideas with others in your organization and earn their support.
- **Start where you are.** Understand the difference between personal power and position power, and leverage your expertise.
- **Surround yourself.** Enlist the help of mentors, a network, and a board of directors to guide your decisions along the way.
- **Know your audience.** Learn which communication channels work best for the people you need to persuade, and practice your presentation, writing, and speaking skills.
- **Rehearse your story.** You will need to have factual information and be able to create and deliver messages that will build confidence and inspire action.

Many of those we interviewed spoke about deciding how they wanted to be known, standing up for their values, building positive reputations, using the leadership styles they embraced, and seeking help to keep them on track. These conscious decisions helped them continue to learn and grow, and to reach new heights.

> Ann Jones is one of those leaders. She felt that her formal education helped her learn about the fundamentals of leadership—where it starts and what it requires. Studying leadership as a subject on its own helped her open up, understand herself better, and realize her needs, values, and vision for the future. As she puts it, "Only by understanding myself can I lead others."
>
> *(continued on next page)*

(continued from page 23)

Soon after graduation, she was promoted to higher levels of management with a major retail company, and her new role brought up challenges immediately. Her peers, some of whom had wanted the same position, suddenly became her direct reports. She talked with them one on one and said she agreed that the new situation was awkward. By creating an environment of openness, understanding, and respect, she reduced the friction for her team members, who responded positively.

Her style is based on respect, involvement, and opportunity for growth. She believes that constant new challenges, recognition, communication, and feedback are critical to getting the best results from her team.

Myth 14: Executives Are Hard to Reach

To gather support within your organization, you will need to talk with its executives. They can be an important audience for your proposals and can become your mentors and champions—or recommend others for those roles. If you believe that executives are hard to reach, this can be a daunting task.

In our experience and that of the people we interviewed, however, this belief turned out to be another myth. Top people in vital, competitive organizations welcome initiative, and many are very approachable. Good executives want to hear new ideas, and they want to mentor and share their knowledge and experience. Here are some examples of the benefits that can follow:

Orrin Matthews wanted to ask the famous CEO of his firm to be his mentor, but he was anxious. This busy man, with a tremendously demanding travel schedule, might not have time for him. After a professor encouraged and guided him on choosing an approach, Orrin mustered the courage to ask. He was astounded when the CEO responded positively, cheering him for his courage and scheduling a meeting to get to know him better within just two weeks. Orrin was flabbergasted and couldn't believe he would be working with the best possible mentor.

Bea Ellison developed a strong belief that her company could introduce more new products faster if it opened its design process. She developed a case and went about meeting with executives to discuss her proposal and seek their insights. Bea was surprised to see how receptive the executives were and welcomed their inputs. They, of course, were delighted with Bea's initiative, and soon she was leading a major project, which was visible throughout the organization. "Be careful what you ask for," Bea says, "I've learned that most often you get it."

Myth 15: Women Can't Flourish in Male-Dominated Professions

We interviewed many women for this book. They are clearly making significant progress in their chosen fields and as leaders. They agree that technical professions have historically been mostly male, but they don't consider the past to determine their futures. To them, it was never a challenge to be in a male-dominated profession. Some of them enjoyed that fact.

Women are still underrepresented in engineering and science, but it is not hard for women to enter these fields. Statistics from the National Science Foundation show that women constituted 40% of those with science and engineering (S&E) degrees in 2005, but their proportion is smaller in most S&E occupations. However, more women than men have entered the S&E workforce over recent decades; their proportion in the S&E occupations rose from 12% in 1980 to 27% in 2007. Women in the S&E workforce are younger than men on average, suggesting that larger proportions of men than women may retire in the near future, shifting the balance.

According to a 2003 Department of Education study, women held only 8.5% of full-time faculty and instructional positions in engineering at degree-granting institutions nationwide. In the physical sciences, 17.2% of such positions were held by women; in computer science, the numbers were better but still not great—30.6% for women overall, 5.6% for women of color. In another study published by the *MIT Technology Review* (2008), an estimated 3,000 PhD-trained female scientists leave the scientific workforce every year, an attrition that costs more than a billion dollars in lost productivity.

Clearly, the factors that contribute to women being less attracted to scientific and technical disciplines, and the obstacles they face when they move forward in their careers, need greater attention and action. Fortunately, a potential solution is already within reach.

More young women than young men are attending college these days. In some areas of the country, the ratio is 54% to 46%. Yet women's rates of participation in science, technology, engineering, and math (STEM) disciplines remain disproportionately low. If colleges can boost participation by women and minorities in STEM-related fields, it will address our workforce needs, as well as the fundamental goals of equity and diversity.

Described by *Time* magazine in 2005 as "perhaps the ultimate role model for women in science," Shirley Jackson achieved many firsts in her career. In 1973, she completed a doctoral degree in physics from the Massachusetts Institute of Technology (MIT), becoming the first African American woman to earn a doctorate of any kind from MIT. Jackson was the first African American to sit on, and then chair, the Nuclear Regulatory Commission (NRC). She also was the first African American woman to be elected to

(continued on next page)

(continued from page 25)

the National Academy of Engineering and to preside over a major national research university, Rensselaer Polytechnic Institute (RPI).

Although she is proud of her groundbreaking achievements, Jackson prefers to focus on her track record in public policy and as an advocate for science and education. She speaks about our need for investment in basic scientific research and for other scientists to become actively engaged in public policy. She recently told a gathering at Harvard's Kennedy School of Government that the exponential rise in the volume and availability of information influences the perception of science and scientists' roles. According to Jackson, scientists must exert consistent leadership to counter confusion over science and mistrust of their work.

Jackson is concerned about impending retirements in science, technology, engineering, and mathematics in both academia and industry; not enough students are in line to replace the record number of retirements coming in these fields. She notes that our national economy and security depend on our capacity for innovation—and that our numbers of scientists, engineers, and mathematicians will dwindle over the next decade unless the trend is reversed.

To counteract this quiet crisis, she says, we must engage everyone, including women and minorities, who have traditionally been underrepresented in STEM fields. Jackson says the crisis is quiet because it takes decades to educate future scientists and engineers, so the impact unfolds gradually. "Without innovation," she says, "we fail as a nation and as a world." She reasons that the ebbs and flows in science funding across disciplines have a "deleterious impact on the creation of a new generation of scientists and engineers"—therefore, our innovative capacity against a backdrop of increasing capabilities abroad.

Nationally, about 25% of graduate students earning advanced degrees are women. They find the leadership component of their studies to be valuable as they pursue their academic and professional goals. These studies help them develop the mindset and skills to be their authentic selves. We expect the numbers of women in technical fields and leadership roles to increase in the coming years.

Chapter 3 Reflection Questions

1. How often are engineers and scientists in your organization promoted to management?
2. How do engineers and scientists demonstrate their leadership capabilities?
3. How are you taking steps to exercise your influence as a leader?
4. What kind of impact do you want to have on your team and your organization?

5. How are scientists and engineers regarded in your organization?
6. How are scientists and engineers involved in business decisions or future strategies?
7. Are women adequately represented in your organization's leadership ranks?
8. Who are some of the women who have succeeded, and how have they done so?
9. Does your organization have a policy for flexible employment? Do employees take advantage of it? What are the perceived benefits and drawbacks?

Societal and Family Beliefs

Myth 16: My Parents Always Warned Me . . .

Your behavior is driven by your beliefs. Many of these you learned in childhood from your parents and were reinforced by the other major influences in your life: school, religion, friends, and media. These people, events, and circumstances shaped you in both positive and negative ways. You may remember things these powerful people said to you or about you that have stuck with you for life, shaping your mindset about yourself and the world.

But even the best advice can have negative consequences. With your young mind, you may have misinterpreted the guidance you received and formed beliefs that have not served you well. For example, when you were told to be modest, you may have heard that you should not speak your mind. The warning to not do dangerous things might have become a fear of taking any chances at all.

You may not have been encouraged to explore, to be creative, and to be adventurous if the adults in your life felt these things were too risky. These beliefs might have stifled you and kept you from being yourself, making a difference, and doing what you really enjoy.

Here's another unintentional result of early influences. If your parents or grandparents endured hard economic times, they may have pushed you toward particular formal education and steady employment, regardless of your true interests or talents. Or they may have encouraged you to develop useful skills you could "fall back on" if your other efforts failed. You carry the deep beliefs that you learned very early in life, even if you are unaware of them today.

To examine your beliefs, start by figuring out where they began. Think about things you do without understanding why. Here's a simple exercise for thinking back on the past.

On a large sheet of paper, draw a horizontal "life line," which represents the time from your birth to the present. Recall the times you enjoyed the most—what they were and what characterized them. Note what happened at those times. Then do the same for the times you enjoyed least. Mark the most positive events on points above the line, with the most negative ones below.

Once your life line is filled with peaks and valleys, consider what patterns you formed as a result of your experiences. Think about what you learned during and after each major event. This should give you some insight into the beliefs you have held, as well as your behaviors and actions. Identify your seven most deeply held beliefs, and consider how they inform your life today.

From these reflections on the past, you should be able to look to the future with clearer vision. Think about how you can be more intentional as you move forward. Decide which beliefs have served you well enough to keep, and which you can let go. One of the beliefs most people share is that our money shows our personal worth. To increase your financial wealth, you can earn more, spend less, or both. Such a notion of success drives many people to make more, spend more, and own more, with no end in sight. This keeps people in jobs they don't like, afraid to risk change but unhappy to stay where they are. But the best things in life often are not things. Even in a culture like ours, largely driven by consumption, experiences can be more valuable to us and to others than what we possess. By analyzing the beliefs that have a grip on us, we can sometimes break free to create a new set of beliefs that are far more liberating, energizing, and aligned with our personal definitions of success. Here are two examples:

Carlos Juarez pursued continuing education, expecting that, by earning a graduate degree, he would be promoted to a higher-level job in his organization—one where he would be more in charge of his destiny and be able to better provide for his family in ways that he dreamed. During five years of study and reflection, he realized more about himself, his life and legacy, and what his inner voice was telling him. He concluded that his goal was not to be promoted in the company where he was but to fulfill his dream—to go back to Venezuela and teach.

Carlos did not make this decision lightly. He conferred with his mentors, spoke with professors, and checked out programs where he might earn his doctorate. He talked with his family and friends, sharing his deeply held beliefs about what was really important to him. He carefully thought through what he wanted his legacy to be, and it was making a difference in people's lives back in the country he came from long ago. It was a breakthrough for him; when he shared with us, he was smiling from ear to ear. He was free to be his true self—no longer in the game he felt he had to play in order to be successful.

Another who made definitive choices around her career as an entrepreneur is Georgia Clifton. After several years of working in a medical device company, she decided to start her own enterprise with a colleague. They worked night and day to get their business set up. She carried the load for defining the business, getting the funding, and so on. Her family was suffering from her absence; she felt torn between home and work. After two years of living such a divided life, she decided the most important thing in her life was to take time off to be with her family.

She, too, had spent long hours in reflection about what was "success" for her. She didn't want to have regrets later in life. She chose to leave, while planning to return someday; she still wanted to pursue the business. At that point in her life, however, the most important thing to her was family time. She was relieved after making her decision. She knew it was right for her and that she would have plenty of time and space in the future to live out her entrepreneurial dreams.

Consider the story about Jerry Johnson in Chapter 1. He says today that his "layoff" was the best gift he could have received. It forced him to find a calling that was true to his values. Today he measures his success by the difference he makes directly in other people's lives. His family now joins him in his work. They have all made clear decisions about family needs, budgets, and savings, which will sustain them while they pursue what is important to them.

Myth 17: We Believe the Baggage We Carry from Our Past Experiences

We get lots of advice from others that reinforce some myths and create new ones. Often well-meaning but uninformed, this advice comes from school counselors, teachers, parents, friends, and the media. Think about how often you have heard advice like "women shouldn't go into engineering" or "it's unusual for men to become nurses." High school counselors advise students not to attend postsecondary technical school, unaware that for many this is a preferred route and that there is a desperate need for more middle-skilled employees. Teachers make judgments about individual students at early stages in their schooling that are often based on societal conventional wisdom rather than the interest of the student.

Individual students are seldom asked what their passions are, much less encouraged to pursue them. Adults probably asked what you wanted to be when you grew up, but they probably didn't tell you much about the options available or the choices you would need to make. If you have never taken the time to reflect on what really pleases you, what captures your imagination and keeps you engaged for hours, do it now. See what comes to mind when you think about the joys of your life. Whatever inspires you could lead to a host of new opportunities. It has for others.

Remember what kind of advice you were given, and by whom, that influenced your choices. Also look back at any choices you regret—and when you realized what your true passions are.

Sarah Stevens tells of her experience of being told by her high school counselor that she could never be a scientist. She would never make it, the counselor said, because the rigor of study would be more than she could handle. She was crushed. It was her dream to work in a laboratory and help find cures for diseases that cripple so many lives and kill people unnecessarily. She chose instead to become a teacher, much more in keeping with her counselor's advice. By the end of her third year after earning straight A's in education, she had decided she could handle the rigor of science and changed her path. She enrolled at Duke University and pursued her original goal. Since then, she has fulfilled her dream and is a research scientist at the National Cancer Institute. Sarah rejoices over finding her own truth and being freed from the baggage of well-intended advice.

Myth 18: Leaders Get MBAs, Not Technical Degrees

This may have been true in the past, because many people who made organizational decisions believed that MBA training focused on general management outcomes. However, as more educational institutions offer leadership courses for their technical people, and more organizations expect their technical people to step up to leadership roles, the relevance of this belief is fading quickly.

We spoke with executives from organizations that sent their engineers and technical people for continued graduate training. We asked what they would value most when college courses were created or updated. Executives at five major firms said they wanted their technical professionals to be better equipped for general leadership. They expected their R&D people, engineers, and technicians to be more closely involved with customers and to bring a broader range of skills to cross-functional business teams. They said, "Please add leadership courses to your curriculum, and ensure that leadership content is integrated in the technical courses." We responded with new course content, and the students and organizations alike are pleased with the results.

This is an early beginning. In organizations and in leadership literature, the MBA is still considered a direct ticket to leadership. For that matter, a leader with both a technical degree and an MBA will be recognized more—or only—for the business degree.

A business education helps technical people gain broader perspectives, learn how to look at systems, and understand the need for other specialized skills. It helps them recognize others' needs and see the importance of delegating responsibilities. We have seen instances where an MBA is the only route to professional advancement, and where sales experience provided valuable career management skills. However, without a basis in technical knowledge, these added abilities provide a poor foundation for growth. Today more engineering and science institutions see the value of a broader curriculum, adding business-related courses and encouraging graduate students to take elective courses from business colleges. We hope this represents a trend toward more frequent integration across disciplines, as well as greater latitude in designing graduate degrees, which would make higher education more relevant and meaningful for everyone involved.

In a presentation about the needs of the 21st century, Dr. Joseph Bordogna, former deputy director of the National Science Foundation, quoted from a 1962 paper by the Institute of Electrical and Electronic Engineers (IEEE) Fellow Maurice Ponte, titled "A Day in the Life of a Student in the Year 2012 AD." As Dr. Bordogna pointed out, Ponte predicted that miniature algebraic computers would replace slide rules and that students would receive satellite transmission of engineering courses.

Dr. Bordogna then mentioned a follow-up publication from 1999 by IEEE Fellows Edward Lee and David Messerschmitt, titled "A Highest Education in the Year 2049." Their vision, like that of Ponte many years earlier, is striking: cyber universities and artificial universes, enabled by high-definition, three-dimensional telepresence; courses in network ontology; and software linguistics. With the new technologies we have seen since 1962, these seem entirely possible.

The future will continue to bring rapid change, for the same reason we have seen so much change in the past century: imagination. This characteristic drives engineers, scientists, and artists alike to see what else can be discovered or created. Imagination, which Albert Einstein said was more important than knowledge, is at the core of all innovation.

Technology leaders of the future, beginning now, need more than first-rate technical and scientific skills. They need critical thinking skills to make the right decisions about the use of resources—time, materials, money, and human effort—in pursuing a goal. Dr. Bordogna described people with imagination in a fresh and interesting way: "They are never confined by what they know, never restricted by existing rules, and never afraid to propose what no one else had seen or imagined. They swing with no net but never lose sight of the ground."

The most effective way to inspire imaginations and keep them active is through education—the foundation of all capabilities, human and institutional. Our understanding of the learning process is the key to unlocking individual potential, empowering a global workforce, and maintaining our democracy. In a world increasingly defined by science and technology, technical professionals must become leaders in their own fields and step forward to lead others.

Myth 19: Engineers and Scientists Are Shy and Keep to Themselves

Engineers and scientists have a shared reputation as introverts, shy types who keep to themselves and their tools. Even when this appears to be true, it does not preclude leadership development. When the analytical side of the brain has been well developed and reinforced, the creative side still has potential. People who have learned to think logically can also use their natural curiosity to imagine what does not yet exist but may be possible. This reserve of intellectual capacity and creative energy can be a valuable asset to any organization, and it may already be there in the minds of engineers and scientists, waiting to be asked for ideas.

Specialized tools, like specialized knowledge, have value to those who possess them. From outside a field of technical expertise, the instruments of science and engineering may look like gadgets or even toys. But for technical professionals they are the means of interacting with the physical world in ways they can measure and discuss. Devices with predictable behavior help them test their assumptions and draw more reliable conclusions.

The ultimate goal of any tool is to help people do new, different, and better things, which they couldn't achieve otherwise. Discoveries made by scientists form the basis for the technologies developed by engineers, who create new possibilities and extend the human experience. As the power of our new knowledge brings us great opportunities, it also brings tremendous responsibilities. Society needs the talents of the technology leader.

Myth 20: People Want to Keep Their Expertise Secret

The technical knowledge service Teltech was launched in the 1980s, and it relied on the expertise of top scientists, engineers, and technologists. They came from universities, private industries, and consulting organizations across the country. When prospective clients heard they would have access to these experts, they expressed doubt. "Why would they share their knowledge with us?" these early customers asked. "They would be giving away their secrets." These people did not yet understand that technical professionals love to share their knowledge.

People in general enjoy talking about their passions, from fishing to cooking. The same is true for interests such as thermal science and environmental engineering. Those who are experts in any field rarely have secrets or treat their knowledge as something to be guarded. Most of the time, technical professionals will gladly offer their wisdom, opinions, and experience. If you work in science, engineering, math, or technology, you have probably seen and done this many times yourself. Sometimes, scientific knowledge makes all the difference in the world, as can be seen from the following example.

John Abraham, a professor of mechanical engineering at the University of St. Thomas, challenged assertions by Christopher Monckton, a journalist and politician who contradicts most scientists about the nature of global climate change. John was able to build his case with evidence from experts worldwide. He took the leadership role of being the champion for this initiative, acting as a catalyst for the many environmental experts whose work he cited. These experts and others rallied to document a comprehensive case that supports their collective work. John's leadership and courage helped them work together. This group of scientists and engineers now provide a "rapid response team" on climate issues for the news media, taking the initiative to clear up misconceptions quickly.

Emerging leaders know that, to succeed, they must share knowledge and collaborate across disciplines to build strong organizations that can survive in a demanding marketplace. With the help of enlightened leaders, organizations can break down barriers between functions and levels to build environments where people realize their interdependence. Technical professionals are willing to share, to do their part. And every day more of them are ready to lead.

Chapter 4 Reflection Questions

1. What myths or misguided beliefs are you ready to leave behind?
2. What new beliefs will be your anchors as you prepare for your future?
3. As you think through your talents, passions, and values, what stands out?
4. Can you align those personal characteristics with your notion of success?
5. Does your work involve your imagination now? If not, how can you engage it?
6. What lessons have taught you something new and valuable?
7. Where do you want to stretch yourself with new beginnings?
8. Create a life line, as discussed at the beginning of this chapter, and identify your peaks and valleys. What patterns and habits did you form during and after these experiences?
9. What do you want to change in your work, your life, and the world?

Summary—Exploding the Myths

Myths are like gossip. Even though they are usually unfounded, they can be dangerous, damaging, and persistent. They get in the way of clear thinking. By examining your beliefs, you can determine which ones remain real for you and which are really myths.

Open inquiry and sound information will lead you to durable knowledge and solid judgments, all of which you will need to address the challenges ahead. Not everything can be proven, and you can't always have every detail you want to build a perfect case. But don't let the perfect become the enemy of the good.

Now that you've awakened to the basis of your beliefs, it's time to begin the real challenge—finding your true self. The next chapter will lead you through a process to do just that.

PART 2

Finding Your Inner Leader

This section invites you to look very closely at yourself as a starting place for your leadership journey. It investigates the following questions:

- What do you truly know about who you are?
- What is the truth about your beliefs, your strengths, your possibilities, and your leadership potential?
- How have you come to know this about yourself?
- Who have been your teachers, mentors, and supporters who have helped you know that you are competent, capable, and ready for leadership?
- How might you reconsider your dreams and the possibilities for your future?
- How do you continue to grow your self-awareness, as part of your ongoing growth and development process—the leader's journey?
- From whom and how do you seek support to help you stay on your path of learning and leading?

This section requires that you undergo plenty of self-reflection and undertake some serious activity to start on the first steps of your own roadmap ahead. ●

The Truth about You

The reflection questions from Part 1 should have you thinking seriously about your beliefs—an excellent starting point for finding your true self. Your beliefs form the perspective from which you see possibilities. Identify which beliefs remain true for you, which you may need to leave behind, and which new ones you want to adopt because you discover they have deep meaning for you.

Behavior, which is visible, often reveals our beliefs, which are not. Make a list of your personal habits, and ask friends and family members to write down the habits they observe in you. From these lists, you can identify what beliefs are driving your actions. Some will make sense; others may be harder to connect. Think about five influences in your life—family, friends, religion, the media, and school—and see if you can track down the underlying belief and its sources. You may find some surprising insights.

> Until late in my thirties, I was a pessimist, yet things had gone very well for me and pessimism didn't fit. I began to ask where it came from and traced it back to my grandmother. Realizing the source of this belief and its inappropriateness, I set out to get rid of it. It took a year or two, but I am now totally cured. In fact, I am now an incurable optimist, and things are still going well. This works for me.
>
> —Author RJB

It's useful to identify a set of beliefs that you clearly know to be *your* truth— what drives your daily thoughts, desires, and actions. This becomes your story. You get to create your own story, so, if you discover beliefs you do not like or want to keep, you can choose others that really represent what you want to believe, and how you want to live.

> Most of our interviews revealed deeply held beliefs. Keith Kutler had beliefs in hard work, compassion, and care as his leadership core. He recognized that these beliefs came from his parents and how he was taught. It worked, and continues to work, for him. Orrin Matthew's story reflects a core belief that, when people are treated fairly, they will do their best to serve. You can count on them. He saw that this belief came from his own experience,
>
> *(continued on next page)*

(continued from page 39)

thinks of it as his truth, and practices it as a leader. Ashley Smith believes that being a good leader simply means being a good person—someone others trust to act with the best intentions. One of her core beliefs is to live in integrity. People around her know they can count on her support and help; she will always deliver on her promises.

If you listen carefully to others' stories, you begin to hear their core beliefs. Practice paying closer attention to what's underneath their statements. See if those beliefs are consistent with their practices and behaviors. And notice your own stories and behaviors. Are they aligned with your beliefs?

Another way to begin discovering your truth is to write your own history. The lifeline exercise discussed in Chapter 3 can help you get started. If you wrote your autobiography, what would the chapter titles be? This exercise begins to show how you became who you are—your beliefs, the experiences that have taught you, and the transformational learnings that shaped your direction.

The exercise of writing my own autobiography as a graduate student was a profound experience that helped me to understand deeply the imprint of my family, the deep beliefs my parents handed down, and what major life experiences shaped my unfolding life, my beliefs and career choices. Since that initial experience, I have continued to write the second and third chapters of my unfolding story. I have come to realize that I get to shape the story for myself. It doesn't have to be totally the imprint of others' impact on me. I get to choose those who are in my influence circles and how they support my unfolding story. The experience has served to keep me focused on the intentional choices I truly want to be making as a leader and contributor in the world at large.

—Author ERM

One more approach to understanding yourself better is self-assessment, using tools that help you identify your strengths, talents, and gifts. Start by listing strengths you know you have and others have recognized. Consider using the Clifton StrengthsFinder, www.strengthsfinder.com/home.aspx, based on work published in Buckingham and Clifton (2001), *Now Discover Your Strengths*. This simple online tool can profile your top talents out of a list of 34 talents identified by the original authors of this research. You may also learn about signature strengths or character strengths in Dr. Martin Seligman's book *Authentic Happiness* (2003) or at www.authentichappiness.sas.upenn.edu.

These tools can help you identify, name, and claim your talents and strengths. The next step is to determine how you use these strengths in your life and work. Are they named and claimed in your actions, behaviors, and beliefs? How do you ensure that you are putting those strengths to work to serve your organization—and yourself?

Too often, people dwell on their shortcomings and weaknesses. Focusing on the downside seldom works, because studying failure teaches us the characteristics of mistakes—not excellence. To learn about success, we must study successes. This does not mean overlooking weaknesses but leveraging strengths in ways that reduce weaknesses, or make them irrelevant.

Emerging leaders working with us identified their strengths, then used an assessment tool that gathered 360° inputs from those around them. The process was enlightening and revealing. Many found that others had rated them much higher than they had rated themselves. They asked themselves why. Could they not see their strengths? Were they too humble about naming and claiming their strengths? Did they lack confidence in their strengths? What was at play?

Paula Hetherington was anxious about learning what her peers, superiors, and subordinates had to say in her 360° feedback results. When she received her feedback, she was shocked. They had all rated her much higher than she rated herself. She had to step back and think about why the gaps were so apparent and consistent. Could she trust these data? Were they all just trying to make her feel good about herself? In a consultation, Paula wondered why she rated herself so low. Was that truly how she saw herself? She thought about how to see herself in a different way, and how to respond to her raters.

It took a while for Paula to accept the feedback she had received from others as the truth. Their open-ended feedback specifically named the strengths and talents they saw in her. She had to own that as her new truth.

George Paulson also rated himself low and received his feedback with trepidation. Once he reviewed his results, he felt amazed that his raters were so generous in their perceptions of his capabilities and performance. To accept their glowing feedback, he had to reframe his perspective about himself and his demonstrated performance in his group. He acknowledged working hard to exhibit the strengths mentioned in his feedback but felt he had often fallen short. Whereas his raters felt his efforts had significant impact, his perspective said "it wasn't enough." The challenge for George was to reconsider what was enough. What were the standards he was holding himself to, and where did they come from? Could he allow himself to see his full capabilities?

This tendency to underrate ourselves is often part of the "high achievement" identity. Good, better, best; one can always be better. This belief can be admirable—or crippling. Do you ever feel that you fall short when rating your own skills or capabilities? Where do those perceptions originate? How might you change them?

Let Your Life Speak

Self-reflection helps us to know ourselves, to determine who we are and who we want to be. It is also important to have conscious goals—to know *why* we want to become different in any way. As Parker Palmer points out in *Let Your Life*

Speak (2002), throughout his career he admired many people. When he tried to be like any of them, however, he became dissatisfied. After careful reflection, he realized that he was trying to be someone else, not himself. It took deep searching inside to determine who he was, what his beliefs were, and who he wanted to be. He wanted to be himself, to be real and authentic. Until he looked inside, he did not know who that was.

We found the same thing with many emerging leaders. They confessed that they had never thought about themselves. They realized that most of their choices had been dictated by guidance from many others around them. They had not taken the time to look carefully at who they truly were. They hadn't assessed their values, their passions, their beliefs, or their strengths in any systematic way. Rather, they had followed a path that seemed comfortable, aligned with their interests, and pleased their close associates and family. They had taken their cues from others or had tried to emulate others. It was rare to hear that someone had truly spent time to be intentional and deliberate in their choice making.

Your life speaks to you. It tells you what you want to become. Some refer to this as a calling or a vocation. When you know yourself, you will have the motivation you need to do the hard work of pursuing your dreams. As a colleague often says, "I've never worked a day in my life, because I've been pursuing my passion." It makes all the difference in the joy we get out of our work and out of living.

The journey to self-awareness is a lifelong process of raising consciousness, reflecting on our choices, and asking ourselves hard questions. Sometimes the answers take a long time to accept. Reflect and act with conscious intent. Create opportunities to stop occasionally and seek guidance from your inner sage. This is how you grow and develop—and reveal new answers.

To help emerging leaders find their inner sages, we use an exercise based on the work of Carl Jung. Better known for writing about the inner child, Jung also studied the inner voice of wisdom. Within each person is a source of wisdom that knows the truth, has learned key lessons from life, and can be called upon.

Students were asked to project themselves forward to an age at which they would be wise and sagelike, having learned the key lessons of life. They would be able to give expert counsel to a young person struggling with questions about life. Once they had chosen that age, they were to put themselves into that sage persona and write a letter to the young person present at this time, offering guidance on the important things to remember about life—the important values that would guide their decisions, some of the keys to living life fully, and what truly matters when all is said and done.

The output of this exercise is always provocative and profound. The learners are amazed to hear their own sage's advice. They are invited to read their letters aloud. When one or two are moved to do so, others follow. The guidance received is often some of the best they could ever imagine. At the end of building their leading and learning plans, students often say this simple exercise opens up rich material that provides a solid framework for imagining their future—as though divine oracles had spoken to them.

Times of growth are beset with difficulties. But these difficulties arise from the profusion of all that is struggling to attain form. Everything is in motion: therefore, if one perseveres, there is a prospect of great success.

—I Ching

To find your inner truth, you should talk about the process of development and change, personally and collectively. Research clearly shows that human consciousness develops through a series of stages, and that those stages are always in the same order. Development follows an invariant sequence in all cultures, as universal and inevitable as nature.

Transformation is the movement from one progressive developmental stage to the next. At each stage, a new "design" principle is used to relate the self to the world. Reality does not change; what changes is the way we organize our relationships to the world. What was unimaginable in a prior stage is suddenly possible. People experience new bursts of creativity, efficacy, freedom, power, and joy. The outer world experiences a person standing more fully as a leader— someone who is capable of greater contributions and service.

Only as the bulk of the population develops to a new stage of development can the whole system take an evolutionary leap. Human development is in the driver's seat. Psychological researchers such as Piaget, Kohlberg, Gilligan, Loevinger, Maslow, Kegan, Hall, Fowler, and Wilbur have described a series of stages we go through, from infancy to the highest stages of adult morality, self-conception, and spiritual consciousness. These and many other theorists, through independent research, have arrived at very similar stage descriptions.

The Transformation to Mastery

In his book *Mastery* (1992), George Leonard describes the transformation process as growth toward mastery. He suggests that transformation follows a learning curve.

Learning anything that requires ongoing practice follows a predictable cycle: a sudden burst or breakthrough to a new level of performance is followed by a small contraction—the inability to fully maintain what was learned. Then there is a long period of seemingly no growth. Leonard refers to this as "hanging out on the plateau." A great deal of learning is actually taking place, but it is not as noticeable as the breakthrough period. The plateau is a time for learning to be digested and incorporated into the structure of the body, mind, and spirit. This time of integration is essential preparation for the next leap forward.

Our leadership development process was deliberately designed to provide gaps between modules, allowing time for emerging leaders to recognize their growth. They also made intentional applications within their workplaces to test themselves, try new behaviors, take on action learning projects, observe themselves as new leaders, seek feedback from those around them, and spend time with their mentors. They alternated between reflection and action. After completing the full learning process, these leaders had clearly moved into new stages of their individual development.

At the beginning of the leadership development process, called Stage 1, we found they had rarely thought about themselves or developed any self-awareness. They expected their learning to result in promotions, either in their own companies or elsewhere. They had little or no awareness of their leadership potential, their emotional intelligence, their learning styles, or their competencies as leaders. Many had very narrow worldviews. They were motivated by earning more money and acquiring new knowledge to further their careers. They knew they were good at manipulating things, designs, or external activity. They saw little value in reflection and were mostly focused on action. They had little exposure to humanities or liberal arts topics, and many were quite naive about their learning agenda or process.

In Stage 2, after doing their first self-assessment and proposing visions for their leadership journeys, they built awareness of their emotional intelligence in the context of other intelligences. They discovered a considerable amount about themselves—their capabilities, learning styles, motivations, passions, sense of potential, and guidance from their stage. They planned courses of action and developed visions of what they wanted to create in their lives, integrating their values, beliefs, and emotions. They saw themselves as beginners in building their worldviews. They understood that leadership is learned, and that they can begin to act as leaders wherever they show up. They learned how to use self-reflection intentionally. They developed support groups to help them on their journey and were ready to receive guidance from them.

After a couple more years in the leadership process, the group focus moved to Stage 3, which was characterized by better understanding of their impact on their teams and organizations. By this time, they had developed a clear sense of their plans. They knew how to reflect on their learning progress, their growth as leaders, their idiosyncrasies, how to move out of their comfort zones, and how to persuade and communicate effectively. They had begun to test their abilities in the workplace and practiced living with conscious intent and authenticity. They appreciated their abilities to work in team settings, leading and influencing outward, downward, and upward. They had become eager to learn, rather than just to complete classes. They saw value in the learning process. They discovered new possibilities for acting as leaders—taking risks and speaking up.

At Stage 4, after the final module on global perspective and action, they demonstrated an expanded worldview—seeing themselves as making a difference that was aligned with their new vision of leadership and impact. They looked for ways to continue expanding that vision and acknowledged their significant growth. They had learned how to change themselves and influence their teams, organizations, and communities. They had a broader view of effective leadership, knowing how to continue testing their capabilities in order to make a difference. They reshaped their plans for the next 5–10 years, often using more leadership actions in their communities, schools, and social circles, as well as their business organizations. They were more actively involved in helping others develop. They were better able to see the intersection of inner work (reflection) with outer work (actions in the world) as a key to growth. They had new ideas of what a masterful leader looks like through readings, experiences, and observations. They accepted the challenge to think provocatively about the world, the issues and the challenges the world faces, and their role in making a difference.

The interview stories in this book show where they are now, years later, what they have learned and integrated, what is important to them today, and how they influence others in their sphere of actions. They are living and acting from a place of new integrity and awareness of their learning and leading agenda for years to come.

Chapter 5 Reflection Questions

1. How do your deeply held beliefs inform you as you attempt to act more intentionally? How do you react when you discover obsolete beliefs linked to some of your actions?
2. When you think about the transformational breakthroughs in your life, what stands out for you? What got expanded? How do you know?
3. Have you ever taken the opportunity to do some reflective "time out" from the work world? What happened for you as you deliberately took this time?
4. What do you know about your truth at this important point in your life? What is the emerging edge of your growth and development?

6

Assessing Your Leadership Potential

Although some people see themselves as leaders early in their lives, most do not, yet many have untapped or unrecognized potential. It is often only with hindsight that we recognize evidence of our leadership capability long ago—possibly in elementary school, in family circles, on the playground, or in sports. Think about where your inner leader showed up in your past, and about who may have recognized it then and encouraged you to step further into your capabilities.

In our research, we have found that leadership is often more visible from the outside—that others may recognize leaders before they see themselves that way. These outsiders recognize something that needs to be stimulated and encouraged. This was confirmed through some of the individual 360° feedback assessments with peers, supervisors, and direct reports.

Do you recognize early leadership traits when you reflect on your past? If not, why do you suppose you never recognized them? Have you ever been given new duties, new responsibilities, or a promotion? Were you told why you were chosen to lead? If you're like most of our alumni, you had to figure it out for yourself. Maybe no one ever explicitly said what they saw in you. Learn to look for the signs that others notice your leadership potential; then find out what it is about you that inspires their confidence. For many of the technical professionals we studied, it is personal integrity, perseverance, knowledge of how to build healthy relationships, or a history of getting things done. The stories of self-discovery are as varied as the people who lived them. Read on and see if you recognize yourself in any of these:

> Dan Jansen recalls recognizing his leadership ability when he gathered a group of colleagues and proposed changing the way they approached the next major aircraft development. Because he felt it was the right thing to do, he didn't ask management's permission. He views this as a result of his own "momentum" coming off a previous project that did not go well, and he wanted to improve the process.

Raymond Adams first recognized his leadership ability as a working adult graduate student. As a young engineer, he didn't have the tools or confidence to take on leadership roles. As he learned about leadership and developed confidence, he began to understand that he had a systems perspective and good critical thinking skills, and he understood how he added value to the organization.

Corrine Anderson always liked being "in charge" as a youngster but shied away from leadership roles as a young professional. When she became a manager, she found that suddenly she was in contact with other people of influence and was in a venue where she could be heard. She always had the courage but just did not feel that people would listen to her. When she confidently began to speak her truth, others were impressed with her ideas and her ability to passionately pursue her vision.

Ashley Smith learned through her leadership pursuits about her strengths and the areas she wanted to enhance. One was her ability to be daring and confident when taking risks. She noticed that when she became apprehensive she tried to analyze her actions and better understand why she was apprehensive. She was not trying to become more of a risk taker but wanted to understand the reasons behind her actions, so that she would be more daring in the future. She also learned that her aspirations had changed. Originally, she wanted to be an executive within a company setting, but through her experience in life and work her goals were no longer position-focused. Now she wants to achieve a "meaningful position" where she can feel she is "making a difference and helping others." These are her goals as a leader. She enjoys being active in her community and volunteering on multiple projects, and she hopes to pursue those avenues more in the future.

Harry Jaspar was one of those introverted, shy types as he began his leadership quest. Even though he learned about many important strengths through his 360° feedback, he was very concerned about his shortcomings when presenting in front of an audience. He was determined to learn how to speak with greater confidence. With coaching encouragement, he joined Toastmasters at his workplace; his peers within the company gave him helpful tips and techniques for overcoming anxiety. He practiced giving short speeches in front of the mirror at home, challenged himself with responding more often in classes, and soon realized he wasn't frightened to speak up. When he gave his final presentation for the graduate program, he received a standing ovation. His classmates gave him great praise for having the courage to strive for this important personal change.

As you assess your potential, think back to the time when you first became aware that you are competent. Was it a particular moment in time, or did you see it as a gradual understanding? Most of us see it as a series of experiences where we reinforced specific skills, knowledge, or capabilities. We probably never thought of the term *competence.*

Moving to Conscious Competence

To show areas of growth from unconscious competence to conscious competence, we frame them in overlapping circles. (See Figure 6.1.) The small circle represents "What I Know," the next larger circle is "What I Know That I Don't Know," and the largest circle is "What I Don't Know That I Don't Know." At the beginning of the learning process, we unconsciously believe we "know so much." The journey toward conscious competence helps us realize how much more there is to learn.

"As I get older and wiser, I realize that I don't get wiser as I get older."

—Mrs. Jalkio, mother of Professor Jeff Jalkio, University of St. Thomas

Figure 6.1 helps us do three things: remember that learning means growth, remain conscious of our levels of competence, and build awareness of our blind spots—where we don't know what we don't know. Of course, we can't know or be competent at everything, and we don't need to be. But we do need to know where we are competent, and we can improve only if we know why.

As we pursue more about what we know we don't know, we continually expand the circle of what we know—and as we become more aware of what we

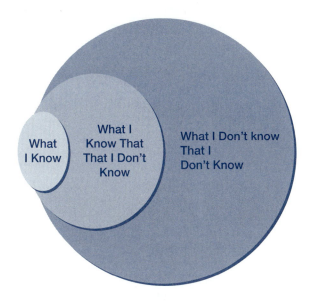

FIGURE 6.1 Moving to conscious competence.

don't know that we don't know, we push the edges of our knowledge and competence. We learn to see where we once had blind spots, and we rediscover how much more there is to know.

Conscious competence leads to confidence, the foundation for courage. To gain that courage, we need to become aware of the basis and extent of our competence. To become better at anything, you need to understand why you are good at it, and what shortcomings you need to overcome in order to increase your competence.

When hiring salespeople, one author often opted for conscious incompetents, because they understood what they needed to work on and, if they were motivated, could become better. Unconscious competents, on the other hand, may be good but they don't know why, and therefore have difficulty improving.

Are you a conscious competent? When did you realize your competence? Between the realization of competence and a conscious understanding of it, most people need time for reflection and discovery. Some find it immediately, and some accelerate the discovery process through conscious questioning. In what situations did you first recognize your competence, and when did you understand why? Really understand. Give it some thought.

> Dan Jansen had an experience as a helicopter crew chief on whose decisions many lives depended. Although when he first joined the service he had low self-esteem, life-and-death experiences every day forced him to take on leadership duties. This experience taught him the necessity of knowing what is important and what is not. It is critical to be right, and it is essential to put aside all self-doubt. You need to be decisive, say what you will do, and do what you say.

The foundation of good leadership is trust. People will not follow those they do not trust, no matter how technically competent or smart they are or how great their ideas sound. One fundamental element of establishing trust is being vulnerable—that is, willing to accept others' knowledge and ideas—along with the ability to be open and honest about oneself.

As hiring managers, we authors would often interview prospective salespeople for offering a technical knowledge product to the industrial market. Personal integrity was a top priority. One question we asked was, "Tell us about the most difficult situation you've ever been in; how did you handle it?" The objective of the question was to see whether the candidate would talk about something substantive or not. How you respond says volumes about you—about your openness, your reliability, and whether others will trust you.

Some of the responses to that question were trivial, some profound. Compare these responses: One person said the most difficult experience was getting a B grade in a course, which was resolved by going to the professor and offering to do more work to get an A. Another talked about having alcohol and drug problems as a youth, going through intervention by friends and family, and completing treatment that dealt with the fundamental cause. Which person would you trust more?

To build trust, you need to accept others—and yourself. It means letting others in, listening, disclosing, and empathizing. It means exploring, evolving, and standing up for what matters to you and what you want in your life. It means being with others, joining with them, sharing the joys and sorrows of life, and building synergistic relationships. When you have trust, you don't need to defend yourself or pretend to be what you're not. We all wear masks at times to protect ourselves and to impress others. Most people can see through this kind of projection and stand apart from us then, leaving a gap where trust should be.

Who among your admired leaders are known for being trustworthy? How do they behave that allows this trust to be built? How do their organizations respond?

A clear example is of a senior leader named Jim, the president of a major business unit in a Fortune 50 company. Jim turned the hierarchy of the organizational chart upside down. He felt his role and title should be at the bottom to represent his responsibility to serve and support all his stakeholders—employees, customers, shareholders, the community, and future employees.

Jim shared his true self and his hard-earned knowledge with anyone. Everyone wanted to be associated with him. He was positive, encouraging, open, and unafraid to share what he believed needed attention for the organization to thrive. He focused his attention on ensuring everyone was informed about the truth of the business situation at all times. He expected others to be true to themselves as well, and he welcomed their input on major business decisions.

Consistent with his positive outlook, Jim focused attention on rewards and recognition—highlighting those at all levels who took action for the sake of their groups and the organization. He made good use of the pronoun *we* in all business discussions. People liked Jim and wanted to be part of his team. His trusting behavior begat more trust. He modeled the way for others to step into their truth and to be willing to be vulnerable.

—Author ERM

This kind of leadership has tremendous impact on the organization at large and its ability to thrive and perform at a remarkable level. People want to collaborate, want to share in the winning, and are willing to give their best efforts.

In building trust, you need to understand your environment, your stakeholders, and your responsibilities—then demonstrate to all of them that you have credibility, a cornerstone in building any relationship. All of this translates to effective leadership. When leaders encourage greater initiative, risk taking, and productivity by demonstrating trust in employees and resolving conflicts on the basis of principles, not positions, they increase their credibility.

Overcoming Your Fears

A primary factor in building credibility and trust is being able to identify and overcome your fears. What are the roots of your fears, anyway? Do you fear for your personal safety or that of your family? Do you fear you'll lose your job? Do you fear there will be international financial crises or conflicts? Probably they have to do with all of these things, usually connected to loss of some sort. Experiencing fear is part of being human. You need to understand how and why your fears arise, so you can deal with them consciously.

Most fears fall into four main categories: mistakes, failure, personal pain, and rejection—all related to loss. As are most people, you are likely to respond to at least one of these fears. But so what? What if you make a mistake? Or fail? Or have pain? Or are rejected? Is this a tragedy? Probably not. Then what is it? It's most likely an inconvenience. Think through your fears and consider "the worst that could happen." It usually turns out to be an inconvenience. We may not like inconveniences, but we can deal with them. So what's to fear?

When we face difficult decisions, we often let irrational or unconscious fears prevent us from doing what we know is right. We need to keep the notion of doing what is right in the forefront and deal with any inconveniences that come along, responding accordingly. As Stephen Covey says, "We can choose how we respond." It is truly not what happens to us in life but how we respond that counts. The experiences of emerging leaders demonstrate that confronting most fears actually leads not to inconvenience but to success.

Jill Chang was sure she would never be able to conquer English. Although she was not shy, she was afraid of standing in front of audiences and speaking. She felt she was misunderstood and would never learn to express herself properly. Jill worked hard to practice English—in her leadership courses, in front of the mirror, with friends who gave her feedback, in the car as she drove, and so on. Still, her progress felt too slow. Finally, between her second and third leadership course, she hired a tutor to help her improve her English. After her final presentation in the third session, her class gave her a standing ovation. She was able to succeed by overcoming her fears. She was proud of that outcome and has built new confidence as a result. She will say now, "I can speak in front of any group now, knowing that my English is not perfect, but I am clearly understood."

Curt Bradford was quick to anger and feared that he couldn't manage it. He often felt strong reactions when peers, superiors, or even friends disagreed with him. He often reacted with sharp or hostile retorts to anything that seemed to challenge his thinking or his work. Realizing that his fears were irrational, he practiced slowing himself down and listening more closely when he felt an angry reaction beginning. He asked for explanations of what had just happened or what had just been said, so that he could hear again the intention beneath the statement. This let him think more carefully about how he wanted to respond. It took time and effort, but he learned not to defend himself against unintentional offenses.

Identify the fears that keep you from being your best. Then look for the beliefs and causes underneath those fears, so you can better understand them and find ways to let go of them.

So far, we have described key elements of assessing leadership potential: through key strengths or competencies, with openness and vulnerability that help build credibility and trust, and by identifying and overcoming fears that keep us from being our best. These are only a few of many elements you can consider.

Bill George, in *True North* (2007), describes the process of discovering our potential as "peeling back the onion." He starts with identifying our outer layers, such as appearances, leadership style, body language, and attire. From there, he identifies inner layers that drive and motivate us, including shadow sides, blind spots, and vulnerabilities. George includes some useful reflection questions and exercises to help deepen your assessment.

Our leadership courses have made extensive use of five assessment tools:

- Using 360° feedback, mentioned earlier
- Learning about your preferences based on personality types
- Measuring your emotional intelligence
- Finding and classifying your learning style
- Identifying your values and passions

Each of these tools can add to your understanding of your present situation and provide a starting point for the future you want to create.

360° Feedback

The 360° feedback technique involves responses to open-ended questions, asked of people with every kind of business relationship to the person being assessed. That includes peers, superiors, subordinates, counterparts at vendor and customer organizations, and so on. By gathering information from multiple sources, this process reveals insights and perceptions that might be harder to see from any single perspective.

Personality Type

This assessment is based on the Myers-Briggs Type Indicator (MBTI) personality inventory, which was designed to make the theory of psychological types described by Carl Jung understandable and useful. According to Jung, seemingly random variation in our behavior is actually orderly, consistent with the distinctive ways people use their perception and judgment.

Jung's typological model regards psychological type as similar to left- or right-handedness: people are born with, or develop, preferred ways of thinking and acting. The MBTI sorts some of these differences into four opposite pairs, with a resulting 16 possible psychological types. None of these types are *better* or *worse* than the others; Briggs and Myers theorized that individuals naturally *prefer* one overall combination of type differences. In the same way that writing with the left hand is hard work for a right-hander, people tend to find using their opposite psychological preferences more difficult, even if they become more proficient with practice and development.

The types are typically referred to by four letters, taken from the following possible pairs:

- Extraversion (E) or Introversion (I)
- Intuition (N) or Sensing (S)
- Feeling (F) or Thinking (T)
- Judging (J) or Perceiving (P)

This assessment helps people better understand their preferences and behaviors. It is also useful for understanding and interacting with others who have different personality types. An online version is available at www.personalitypathways.com/type_inventory.html.

Emotional Intelligence

Another powerful assessment tool measures a potential leader's emotional intelligence. As more and more leadership research suggests, building competence in this arena has huge benefits; in fact, emotional intelligence has far more impact than IQ. The importance of emotions in work settings has been established (Ashforth 2000; Jordan and Troth 2006; Weiss and Cropanzano 1996). Emotional intelligence, a multidimensional construct that links emotion and cognition with the aim of improving human interactions (Mayer, Brackett, and Salovey 1997), has been linked to improved workplace behavior (Aritzeta, Swailes and Senior 2007, and Ashforth 2000) and, in particular, team behavior (Druskat and Wolff 2001) and team performance (Jordan and Troth 2004).

Although several assessment tools are available, we prefer using one that measures the underlying capacities—the basic constructs underlying emotional intelligence as defined by all researchers. The following are those basic constructs:

- Awareness of one's own and others' emotions
- Emotional facilitation
- Emotional understanding
- Management of one's own and others' emotions (Mayer, Bracket and Salovey 1997)

In all models, emotional awareness and management are core abilities. Almost all leadership competencies depend on managing our relationships with others, which begins with managing ourselves. Dan Goleman and colleagues (Goleman, Boyatzis, and McKee 2002) and others (e.g., Boyatzis, McKee and Johnston 2008) have demonstrated why these are necessary skills and capabilities for effective leaders. This tool provides a profile of competency in empathy, social regulation, self-control, and access of emotions in daily interactions with others, which leads to effective outcomes in relationships.

Technical leaders have come to appreciate the importance of developing their emotional competencies, and many have found their own results enlightening. By developing awareness and empathy, they learn how to better understand others' emotions and show compassion in open and healthy ways.

Learning Styles

One more helpful assessment comes from Kolb's work (Kolb 1981, 2007) on understanding learning styles. Continued learning is critical for ongoing development; knowing your natural tendencies can help make your learning more effective. According to Kolb, the learning cycle requires four processes for learning to occur (see Figure 6.2):

■ **Diverging (concrete, reflective)**—This ability emphasizes the innovative and imaginative approach to doing things. It means you can view concrete situations from many perspectives and you are able to adapt through observation rather than by action. You are interested in people and tend to be feeling-oriented. You probably like such activities as cooperative groups and brainstorming. You are constantly expanding the conversation at hand.

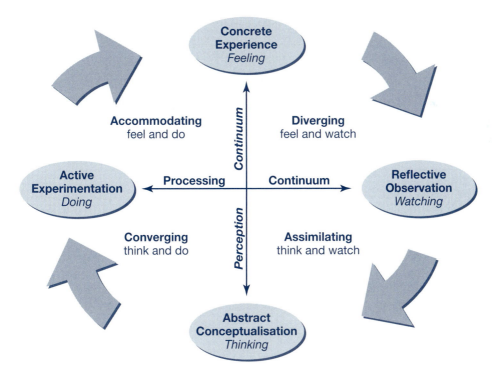

FIGURE 6.2 Kolb learning cycle.
Source: Permission granted by Hay Group, Inc.

- **Assimilating (abstract, reflective)**—This is about the ability to pull together a number of different observations and thoughts into an integrated whole. You no doubt like to reason inductively and create models and theories. You like to design projects and experiments to test what is possible.
- **Converging (abstract, active)**—This ability emphasizes the practical application of ideas and solving problems. You like to make decisions, solve problems, and make practical application of ideas. You prefer technical problems over interpersonal issues.
- **Accommodating (concrete, active)**—This speaks to using trial and error rather than thought and reflection. You are good at adapting to changing circumstances; you solve problems in an intuitive, trial-and-error manner, such as what is done in discovery learning. You also tend to be at ease with people.

Through this assessment, learners can identify their preferences and see how they learn best, creating new learning activities with appropriate designs that incorporate their preferences. This assessment has strong relationships and similarities with other tools and confirms learners' uniqueness and identified strengths. This is equally valuable for leaders as they present information to others.

After reviewing her assessment results, Paula Hetherington said, "I am amazed with what I've learned about myself through all of these assessments. It gives me lots of affirmation in seeing my unique abilities, strengths, and ways of doing things. Now I not only have a much better idea of how to put all of this to work for myself but also I clearly see how this can work for helping my people and our teams better manage how we deliver on our work responsibilities." She was able to integrate all of the assessment results into a coherent picture of her own situation, making it possible for her to better execute her leading and learning plan as a result.

Values and Passions

It is critical to be aware of the underlying values that are not only the guides to better understanding our belief set but also markers for us in our choices in life. They are guides for decision making, provide markers for us as we grow and develop, and help us clarify the most important features in our landscape plans.

Many authors have identified personal and organizational values to consider as you work your leading and learning plan. Brian Hall (1995) provides a broad array of values that encompass all cultures. It is a valuable tool for assessing one's value set.

We authors have developed our own set of values for the discriminating young leader to consider, and it is available in the appendix. Using this tool as a guide, you will need to spend reflective time considering which values serve as your primary drivers for behavior, choices in your life, and important guides in ensuring a good fit with your organization of choice.

Frederick Hudson (1999) has interviewed hundreds of leaders to seek to better understand how core values drive their behaviors and choices. From this compilation of leader interviews, he identifies six core values sets, which he describes as deep sources of passion and commitment:

- Intimacy
- Sense of self
- Achievement
- Creativity and play
- Search for meaning
- Compassion and contribution

The choice of any of these human values is usually a function of where one is in his or her adult life cycle. The selection and balancing of the values make each sequential chapter different from the previous ones, giving each a different sense of purpose and meaning.

Spending time in reflection to identify what brings you great joy and excitement, what are your guiding values, and how you put them to use gives you a clear foundation from which to work. Recognizing how they shift as you grow and develop is fuel for generating energy and life purpose.

For other assessment instruments and their descriptions, see the appendix at the end of this book.

Chapter 6 Reflection Questions

1. What is your leadership potential? How do you know?
2. What assessment have you used in the past? What new assessment instruments may have value for you?
3. What feedback has come to you that suggests others think you have strong leadership potential? How might you gather more insights from others?
4. What, in your opinion, are the fundamentals of becoming a leader? Make a list of those you believe are the most important.

CHAPTER 7

Creating a Vision for What You Want

Now that you have looked closely at your formative influences and assessed your leadership potential, consider what you want to create in your life—and how you see your contribution. For example,

- What kind of leader do you want to be?
- What comes naturally for you?
- What will success look like for you?
- How will you know when you have succeeded?
- What are your dreams for the future?
- How far away is this vision?

To figure out the type of leader you want to be, think of the leaders who have influenced you. What did they do that made a difference? How would you define their leadership styles? Were they demonstrating authentic styles, charismatic, or servant leader behaviors? What were their most admired attributes?

> As we survey the path leadership theory has taken, we spot the wreckage of "trait theory," the "great man theory," and the "situational critique," leadership styles, functional leadership, and finally, leaderless leadership, to say nothing of bureaucratic leadership, charismatic leadership, group-centered leaders, reality-centered leadership, leadership by objective, and so on. The dialectic and reversals of emphases in this area very nearly rival the tortuous twists and turns of child-rearing practices, and one can paraphrase Gertrude Stein by saying "a leader is a follower is a leader."
>
> —*Administrative Science Quarterly*

Leaders come in all sizes, shapes, and dispositions—but most of them have some common characteristics. First is a guiding vision: leaders must have clear ideas of what they want to do, both personally and professionally, and the strength to persist in the face of setbacks and failures. Unless you know where you're going and why, you cannot get there. Leaders invent themselves. They are not born but rather self-made. Some leaders have been made by accident, circumstance, sheer willpower, or perseverance.

Most leadership courses are focused on teaching skills; we have attempted to go beyond the skills and provide a template for helping potential leaders invent themselves in their own way. They put time and energy into studying other leaders and contemplating their own talents and abilities, but in the end they work with their own raw material.

The same is true for you: if knowing yourself and being yourself were as easy to do as saying it, there wouldn't be so many people walking around in imitated postures, spouting secondhand ideas, and trying desperately to fit in.

To set forth your vision, invent yourself your own way. Learn with each step, unlearn when necessary, and become the author of your own life. As Bennis (1989) says, "you can learn anything you want to learn, but true understanding comes from reflecting on your own experience." This takes effort, asking hard questions to reach self-awareness. Nothing is truly yours until you understand it; however, once you understand, you know what to do. Learning and understanding are the keys to self-direction. Our relationships with others teach us about ourselves.

No one can teach you to be yourself. Indeed, however well-intentioned your parents, teachers, and or peers have been, the best they can do is teach you how not to be yourself. As Jean Piaget, the famous Swiss psychiatrist so unequivocally stated, "Every time we teach a child something, we keep him from inventing it himself."

To begin the invention process, you need to commit to innovative learning and creating anew. That's how you realize your vision, exercise your autonomy, and work within your prevailing context in a positive way. It is dynamic and full of possibility, beginning with curiosity and creativity and fueled by knowledge and understanding. It allows you to change the way things are. "We begin to shape life, rather than being shaped by it" (Bennis 1989).

The technical professionals we interviewed charted their courses of action in their own unique ways. One used a Have-Want-Gap analysis. Beginning with the current state and defining the desired future state, this emerging leader determined what had to be done to get from one to the other. First, identify what you have. Consider the more tangible things, such as income, job responsibilities, and home and leisure time, plus less tangible things, such as your sense of happiness, peace of mind, and satisfaction with yourself.

Next, list what you want. Picture an ideal future in which you have the best-case situation for you: a job you enjoy, a sense that you are making a difference in the world, and adequate time to enjoy your family and friends. What are the most satisfying experiences you have had, and what experiences have you not yet had but want in the future?

Having identified your haves and wants, ask two key questions: Why do you want things to be different, and what will it take to create that future?

1. **Why do you want things to be different?** Your primary motivation may be increasing your income, becoming the best at something, making your work more meaningful, making life easier, or reducing your costs. You may be driven by personal goals, such as earning greater respect, increasing your power, achieving more recognition, or becoming more prominent in your group or society. What is it for you?
2. **What will it take to create that future?** If you know what you want, and it's different from what you have, what are the barriers to achieving your

goals? If you know what the barriers are, are you taking steps to overcome them? What is it costing you to not take these steps?

These questions are simple, but the answers are not. Dealing with change is always difficult, particularly if it involves looking inside yourself for your beliefs and the motivations that keep you from growing. Sometimes they are hard to see, and difficult to acknowledge. However, to make progress toward the person and leader you want to be, you need to identify these issues and address them.

When confronting the unknown and faced with the need to change, we are often dealing with four fundamental fears: mistakes, failure, rejection, and pain. We are often our own worst enemies—the greatest obstacles to our progress—when we try to change. Feeling unable to act, even toward goals we want to reach, can be debilitating. Leaders overcome those feelings.

Recognizing Your Potential

Some of the leaders we interviewed did not recognize their own potential before others saw it; the same thing could happen to you. Remember in Chapter 5 how astonished Paula Hetherington was at others' perceptions of her as a leader? Others had similar experiences:

Patsy Hall, a team leader at a major medical device company, was stunned by what she uncovered. She had read self-help books that provided new insights during her time of gathering data and had found the provocative questions in those books helpful in coming to terms with her beliefs, motivations, and passions. She had never asked herself deep questions about her future; she just assumed that, if she did a good job where she was, doors would open for her automatically. She hadn't considered taking charge, making things happen for herself, or seeking guidance from mentors or coaches to help her plan how to achieve her dreams. When all this possibility for asserting herself became clear, she became inspired to move forward and become a person who could shape her future.

Ham Wilson, another employee for a major medical device company, felt that undertaking all of the assessments and reviewing the input from others was beneficial—but he was still skeptical. After his final leadership course, he said, "Well trained to be skeptical of unproven theories and unquantified hypotheses, I approached discussions of the so-called soft skills of leadership with a jaundiced eye. My intractable position was further ensconced by my own unease in discussing, much less practicing, soft skills. Not knowing how to tangibly measure them, I felt ill-equipped to monitor my progress toward developing these skills. Over the past three years I have 'learned how to learn' with regards to assessing and developing these intangible skills. I recognize that they are integral to the context of leadership as a

(continued on next page)

(continued from page 61)

whole, inextricably woven into the cloth from which exceptional leaders are cut. This meta-cognitive lesson is the preeminent skill I carry forward from graduate school, one that opens an entirely new, expanding field of inquiry and exploration for the years to come."

Ralph Schultz, another insightful leader, said, "One of my very wise mentors made a very powerful statement that I will never forget. He said, 'Live each day as you would like to be remembered and love each day as if it were your last.' Whenever I talked to him about my future and who I wanted to become, he asked, 'Where is your heart and is this your passion?' Often, I ask myself the above question to see if I am being true to myself and doing what I know can lead me to become a great leader. Knowing and seeing the big picture, being able to communicate to all levels, treating others with respect, kindness, and empathy, being truthful and honest with oneself are qualities that I see in a great leader. Most of all, I think a great leader's actions, thoughts, and passion should be in harmony with each other."

How do you clarify your vision of the future and get a sense of what is possible? Many of the emerging leaders found their own ways of figuring out how to put this into words and plans that worked for them. What is yours?

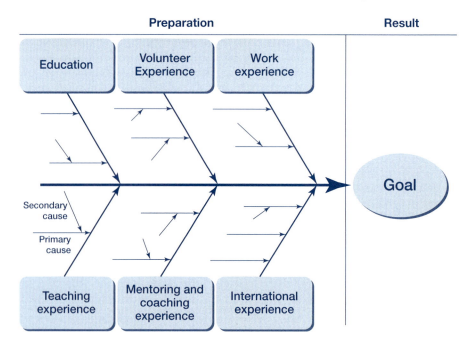

FIGURE 7.1 Modified Ishikawa approach for planning.

Wade Dennison had a long-range vision before age 30. He had a goal of working in developing countries. But how would he get from where he was to where he wanted to be? Wade used a time-tested method in a totally new way. (See Figure 7.1.)

The Ishikawa, or fishbone, diagram is well known to anyone who has worked in quality to track down the source of a problem. The usual method is to put the problem at the "head" of the fish, then list possible causes as "bones" along a central spine. Wade Dennison did just the opposite, putting his long-term goal as the head, then listing all the experiences and skills he would need to get there as the bones. He is well along his route to his goal and has used this model to help steer his education, work, and life experiences to build the credentials he needs to be qualified to undertake his ultimate passion.

Carol Jacobs has a different approach to creating her future. She has no master plan, but searches for new experiences that offer adventure. No matter how seemingly different the new experiences are, she builds on the experiences and networks she has created in the past. This is no "blank sheet of paper" approach to the future, but a thoughtful and strategic use of resources she has developed. Constantly thinking how she can adapt past experiences and personal networks to current situations, she taps into long-standing relationships to help her.

Another leader describes a method he has used several times in his career to help sort out priorities and make decisions in complex career situations with many viable options. Using a spreadsheet, list the options down the first left column. Across the top, list the decision criteria that are important to you. These could be anything: salary, flex time, job satisfaction, location, and so forth. In the row below the criteria, put a weighting number, such as 10 for "high," 0 for "low." In each cell, list the ranking you would give to each option under each criterion. Then, for each option, multiply the weighting by the ranking and record the product. This gives a good first cut at the priority of each option.

Prepare to Think Differently

Regardless of the way you identify and articulate your vision, there are some essentials to consider. Prepare to think differently and expand your horizons. When someone suggests something that seems ridiculous, rather than dismissing the notion, ask the person to elaborate. One former executive of a major medical device company, who had tremendous success during his career, used this approach. His wife says he could spend endless amounts of time listening to a lot of people with crazy ideas. He never put them down, and he always learned something. One definition of visioning says you need to be able to listen to a lot of crazy ideas and keep a straight face.

As educated, critical thinking people, we tend to look for fatal flaws. We are well versed in analysis, but not as much in innovation and creativity. Our left brains are well developed, but our right brains are not. They need more exercise. Organizational life is full of left-brain culture: logical, analytical, technical, controlled thinking is the norm. Right-brained qualities, such as intuition, conceptualization, synthesizing, and creative problem solving, are valuable, too.

In most organizations, managers serve as the left brain, and the R&D group serves as the right brain. A leader must combine both, using what Hesselbein and Goldsmith (2006) calls a whole-brain approach. This includes learning to trust what Emerson called the "blessed impulse," the hunch, the vision that shows you in a flash what is the absolutely right thing to do. Leaders must learn to trust these impulses, to pay attention to the inner voice, which is the purest, truest thing we have.

> Bea Ellison talks about doing the unusual—thinking outside the box, creating and inventing something untested—from hunches she has learned to trust. She takes time for reflection, thinks intentionally, uses keen observation skills, and seizes opportunities to do things that haven't been done before. After a few successes doing things in unusual ways, thinking through new possibilities, she was able to trust her intuition more fully and take more risks. Bea's advice is don't do the same old things or you'll get the same old results.

> Another technical leader shares his experience with brainstorming. He feels you can develop your ability to think differently by engaging others in brainstorming, the open-ended identification of ideas and solutions without restriction. Only after exhausting a list of ideas are they critically evaluated. This process creates an open environment where everyone is encouraged to come up with ideas, no matter how crazy. This process stimulates more ideas and eventually can lead to approaches that would not have been previously considered. Without thinking differently, it is not possible to do things differently. You can take charge and take responsibility for encouraging brainstorming and thinking differently in any group. New and better ideas will emerge.

Many people learn from mistakes, taking risks and sometimes just being terrified. Being in a life-or-death situation can get your attention. Some freeze up and can do nothing. Others rise to the occasion, thinking "What do I have to lose?" and "What do I have to gain?" Most of our daily experiences are not this critical; they do not have the potential for tragedy. But some do. Several of those we interviewed learned from these experiences:

Dan Jansen was in a group in his company that was developing a new device to detect wind shear on runways, a condition that can cause catastrophic failure in landing airplanes. The project failed because the theory on which the device was being designed proved to be incorrect. The concern among the engineering staff was that this failure would dampen executive interest in pursuing other creative projects. However, the team leader handled the failure very well. He wrote a complete analysis of the project, explaining the cause of the failure, and presented it to others in the design group and to executives. Because of this comprehensive analysis and explanation, the company did not overreact and has continued to pursue other innovative projects.

Rae Collins sees marathon running as a metaphor for the time in her career when she was feeling run down and afraid of never reaching her goals. She learned then that she must keep going. She persisted in her work developing a new technology. With the help of her team, she was able to find a solution, pursue a patent, and break the mold for leadership at her company. As with her marathon running, she is proud of her perseverance. She would never have chosen the pain, she says, but recognizes now how it strengthened her and provided her a crucible for learning.

Failure can also be a great teacher. In interviews with 6th-grade girls in a residential engineering summer camp, failure was ranked as the second most fun part of the camp. Early failure of the airplanes they were building led the girls to work together, identifying what went wrong and then fixing it. It was the key to their eventual success. A colleague who teaches neuroscience heard this story and said, "That's what research is all about. I have far more failures than successes in my research."

Mistakes can become positive lessons when we think through them calmly, noticing where we went wrong, mentally revising, and then acting on the revisions. When great golfers miss their intended goal, they don't linger over the miss but, rather, set about to improve their stance or swing. And great golfers do make mistakes. Watch your favorite pro golfer and notice how mistakes have become great lessons for overcoming obstacles with strategic action. The point is to use mistakes creatively.

Reflection Inspires

Like many pioneers, womens' rights activist Gloria Steinem has made a vocation of venturing into uncharted waters. She says, "Too much intellectualizing tends to paralyze us, but true reflection inspires, informs, and ultimately pushes us to resolution" (Bennis 1989). Only when you are able to see your mistakes and failures as a basic and vital part of life can we make conscious that which is unconscious. To do anything well requires knowing what it is that you're doing and you can only know what you're really doing by making the process conscious.

> One author's grad school colleague was in the army ski patrol in the mountains of Italy. His sergeant would constantly tell the team, "If you're not falling down, you're not trying hard enough." We need to learn to fail early and often, then try again. The key is how we respond.

Think of someone you know who has overcome failure, and ask that person some questions:

■ Was the experience good or bad? Why?
■ What did this person learn?
■ What's the upside?
■ Were you encouraged to try again?

In many cases, some failure is essential to getting the next big responsibility. Once you have experienced a failure, you can recognize the signs earlier the next time and take corrective action. If you keep trying, learning from your mistakes, things change in significant ways.

Organizations that disallow mistakes create two types of counterproductive behavior. Mistakes are covered up or tend to be selectively reinterpreted, so that everyone can pretend no mistake was made. Either way, no one learns how to benefit from the experience.

> In sharing his philosophy about mistakes, alumni Ben Kittleson said, "I decided that more than anything else, I wanted to establish a climate where we could encourage people to take risks. We started with the premise that we could accomplish anything we wanted, only if the people were permitted to do what they wanted to do. I soon recognized that it was somewhat naïve on my part to assume that anybody could do anything. However, on the other hand, if you want to believe that growth comes from risk-taking, that you can't grow without it, then it's essential to lead people toward growth to get them to make decisions, and to make mistakes, and learn from those mistakes."

A good example comes from *The Wall Street Journal* (2005) about Warren Buffett, the billionaire investor and insurance executive who runs Berkshire Hathaway Inc. Buffett has created an organization that works well for him and his investors, for whom the company has made estimated annual returns of 31% since 1951, compared to the 11% annual return of the S&P 500.

What happens when someone makes a mistake at Berkshire Hathaway? The company can't generate 31% returns for more than 50 years by allowing people to lose money. Here's the story as reported by the *Journal:*

Mr. David Sokol [chief executive of Berkshire Hathaway's MidAmerican Energy subsidiary] recalls bracing for an August 2004 meeting at which he planned to break the news to Mr. Buffett that the Iowa utility needed to write off about $360 million for a soured zinc project. Mr. Sokol says he was stunned by Mr. Buffett's response: "David, we all make mistakes." Their meeting lasted only 10 minutes. "I would have fired me if I was him," Mr. Sokol says. "If you don't make mistakes, you can't make decisions," Mr. Buffett says. "You can't dwell on them." Mr. Buffett notes that he has made "a lot bigger mistakes" himself than Mr. Sokol did.

Consider Sokol's response: "I would have fired me." Chances are, you'd be fired for a $360 million mistake, too. But that's not the worst part. What's worse is that so many of us fire ourselves from the position of trying anything new. We spend more time kicking ourselves or covering up our errors than trying to learn from them. We fire ourselves from the position of "chief mistake officer," which, on the organizational hierarchy, reports to the "chief innovation officer." We can't have one without the other, so we stop trying.

Finally, putting together your vision requires confidence, taking risks and stepping up. That means when you identify a situation that needs leadership you speak up and make yourself visible. You form a team, clearly identify the problem to be solved, and engage all the skills of your teammates in finding the solution.

This involves risk, of course, but choosing not to step up means the situation will persist. That has risks of its own. Stepping up will require you to demonstrate courage. It can be scary and exhilarating at the same time. It can also be fun. Think of it as embarking on an adventure, or becoming your favorite mystery detective.

People do not often think of themselves as courageous. Most often, we avoid situations that require courage. Sometimes we cannot. So where do you go to find courage? It comes from two main places, both inside you: one is a strong passion for something; the other is confidence.

You are passionate about things that are important to you. Think of the subjects in which you immerse yourself and lose track of time. When you are working with passion, you have the perseverance, persistence, and energy to put in your best effort.

Confidence comes from knowing your skills and abilities—the result of your education, training, and experience. This knowledge has been confirmed by the response of others in the increased responsibilities you've earned.

Validate Your Idea

When you see problems, do you often think you have a better idea of how to solve them? Most of us do. Do you then set out to validate your idea and take action? Most of us don't. Leaders test their inspirations against the opinions of others, building support and refining concepts.

To make the difference you want, define your vision; then validate it by engaging with others. They can encourage you and work with you. Make your personal power known; it is the leadership trait you develop through helping others achieve their goals.

As a concluding remark at the end of his leadership development process, Ham Wilson said, "I believe that crucible experiences are critical not only for validating the foundations from which a leader's authenticity is built; they are also opportunities to flex and enhance a leader's adaptable capacity." Passing through this gauntlet, leaders earn a knowing self-confidence that both acts as a touchstone for future decisions and welcomes future challenges and risks, appreciating the growth potential found even in failure. Bennis describes adaptive capacity as "applied creativity—an almost magical ability to transcend adversity, with all its attendant stresses, and to emerge stronger than before" (Bennis and Thomas 2002). It is the combination of hardiness and ability to grasp context that, above all, allows a person to not only survive an ordeal, but to learn from it, and to emerge stronger, more engaged, and more committed than ever. This inner knowing steels one's commitment, confidence and professional will—while also prompting the Darwinian instinct to learn and adapt, becoming open to the risks associated with change. Adaptive capacity allows leaders to be fully present and aware of changes needed for their organizations in the emerging future, along with the confidence and will to act in an instant.

The following is a final vision exercise, adapted from the Society for Organizational Learning.

Drawing Forth Your Personal Leadership Vision

This exercise begins informally. You sit down and make up a few ideas about your aims, writing them on paper, in a notebook, or on your computer. No one else need ever see them. There is no proper way to answer and no measurable way to win or lose. Playfulness, inventiveness, and spiritedness are all helpful—as if you could again take on the attitudes of the child you once were, who asked similar questions long ago. Pick a locale where you can sit or recline in privacy, a quiet and relaxing space to write, with comfortable furniture and no glaring light or other visual distractions. Play a favorite piece of music, or work in silence if you prefer. Most importantly, give yourself a block of time for this exercise—at least an hour, on a day relatively free of hassle. Hold your phone calls and visitors for the duration.

Step 1: Creating a Result

Bring yourself to a reflective frame of mind. Take a few deep breaths, and let go of any tension as you exhale, so that you are relaxed, comfortable, and centered.

From there, you may move right to the exercise, or you may prefer to ease in by recalling a meaningful image. It could be a favorite spot in nature, real or imagined; an encounter with a valued person; the image of an animal; or an evocative memory of a significant event: anytime you felt something special was happening. Shut your eyes for a moment, and try to stay with that image. Then open your eyes and begin answering the questions in the following paragraph. Imagine achieving a result in your life that you deeply desire—such as where you most wish to live or the relationships you most wish to have. Ignore how possible or impossible this seems. Imagine yourself accepting the full manifestation of this result. Describe in writing or sketch the experience you have imagined, using the present tense, as if it is happening now:

- What does it look like?
- What does it feel like?
- What words would you use to describe it?

Step 2: Reflecting on the First Vision Component

Now pause to consider your answer to the first question. Did you articulate a vision that is close to what you actually want? There may be a variety of reasons you found it hard to do:

"I can't have what I want." Pretending you can have anything you want may not be an easy task. Many people find that it contradicts a habit held since childhood: "Don't think too much about what you want, because you might not get it." In a preemptive strike against disappointment, they denigrate any object of their deep desires. "It'll never live up to expectations, anyway." Or they may feel they have to trade it off against something else: they can have a successful career or a satisfying family life, but not both.

In this exercise, you are trying to learn what your vision is. The question of whether it is possible is irrelevant. That's part of current reality. Suspend your doubts, worries, fears, and concerns about the limits of your future. Write, for the moment, as if real life could live up to your deepest wishes: what would happen then?

"I want what someone else wants." Some people choose their visions based on what they think other people will want for them: a parent, teacher, supervisor, or spouse. For the duration of this exercise, concentrate on what *you* want. You may find yourself articulating that you want a good relationship with (for example) your spouse; you want the time to devote to that relationship, the understanding to act wisely within it, and the ability to live up to the mutual commitments you have made. But you should include it only if you want it for yourself—not because you think your spouse would want it.

"It doesn't matter what I want." Some people assume that what they want is not important. They scribble out whatever comes to mind quickest, just to get "any old vision that sounds good" down on paper. Later, when they need a coherent personal vision as a foundation for further learning, their haste turns out to have been counterproductive. Do not belittle yourself; if, like many of us, you have doubts about whether you deserve rewards, imagine the rewards you would want if you *did* deserve them.

"I already know what I want." During this exercise, you may create a new sense of what you want, especially if you have not asked yourself this question for some time. A personal vision is not a done deal, waiting for you to unearth and decode it. It is something you create, and continue to re-create, throughout your life.

"I am afraid of what I want." Sometimes people say, "Well, what if I didn't want to stay at my job anymore?" Others are afraid that, if they let themselves start wanting things, they'll get out of control or be forced to change their lives.

Since this is *your* vision, it can't "run away" with you; it can only increase your awareness. Nonetheless, we suggest that you set your own limits on this exercise. If a subject frightens you too much, ignore it. However, the fact that you feel uneasy about something may be a clue to potential learning. A year or two from now you may want to go back to that subject—at your discretion.

"I don't know what I want." In *The Empowered Manager* (1991), Peter Block offers an effective approach with people who say they don't have a personal vision ("of greatness," as he calls it) for themselves. In effect, he says not to believe them: "The response to that is to say, 'Suppose you had a vision of greatness: what would it be?' A vision exists within each of us, even if we have not made it explicit or put it into words. Our reluctance to articulate our vision is a measure of our despair and a reluctance to take responsibility for our own lives, our own unit, and our own organization. A vision statement is an expression of hope, and if we have no hope, it is hard to create a vision."

"I know what I want, but I can't have it at my workplace." Some people fear that their personal vision won't be compatible with their organization's attitudes or vision. Even by thinking about it, and bringing these hopes to the surface, they may jeopardize their job and position. This attitude keeps many people from articulating their vision or letting this exercise go very far.

This is really a question of current reality. As such, the perception is worth testing. Occasionally, someone we know does test it, by asking other members of the organization what they really think of this "dangerous" proposed vision. More often than not, the answer is "It's no big deal." When approached directly, organizations tend to be far more accepting of our goals and interests for ourselves than our fears lead us to expect.

Nonetheless, you may be right about your vision's unacceptability. If you can't have it at work at *this* place, then your vision might include finding another place to work that will allow you to grow and flourish. You might also think about your vision for volunteering, working for an organization such as Habitat for Humanity or starting your own nonprofit.

Step 3: Describing Your Personal Leadership Vision

Next answer the following questions. Again, use the present tense, as if it is happening right now. If the categories do not quite fit your needs, feel free to adjust them. Continue until a complete picture of what you want is filled in on the pages.

Imagine achieving the results in your life that you deeply desire. What would they look like? What would they feel like? What words would you use to describe them?

- **Self-image:** If you could be exactly the kind of leader you wanted to be, what qualities would you possess?
- **Tangibles:** What material things would you like to own?
- **Home:** What is your ideal living environment?
- **Health:** What is your desire for health, fitness, athletics, and anything else to do with your body?
- **Relationships:** What types of relationships would you like to have with friends, family, and others?
- **Work:** What is your ideal vocational situation? What impact would you like your efforts to have? How do you want to be able to influence others and your organization?
- **Personal pursuits:** What would you like to pursue in the arena of personal learning, travel, reading, or other activities?
- **Community:** What is your vision for the community or society you live in?
- **Other:** What else, in any other arena of your life, would you like to create?
- **Life purpose:** Imagine that your life has a unique purpose—fulfilled through what you do, your interrelationships, and the way you live. Describe that purpose, as another reflection of your aspirations.
- **Contribution:** What gifts and talents do you know you have that will make a difference in the world?

Step 4: Expanding and Clarifying Your Vision

If you're like most people, the choices you put down are a mixture of selfless and self-centered elements. Part of the purpose of this exercise is to suspend your judgment about what is "worth" desiring and to ask instead which aspect of these visions is closest to your deepest desires. To find out, you expand and clarify each dimension of your vision. In this step, go back through your list of components of your personal leadership vision: include elements of your self-image, tangibles, home, health, relationships, work, personal pursuits, community, life purpose, and anything else.

Ask yourself the following questions about each element before going on to the next one.

If I Could Have It Now, Would I Take It?

Some elements of your vision won't make it past this question. Others pass the test conditionally: "Yes, I want it, but only if . . ." Others pass and are clarified in the process.

People are sometimes imprecise about their desires, even to themselves. You may, for instance, have written that you would like to own a castle. But if someone actually gave you a castle, with its difficulties of upkeep and modernization, your life might change for the worse. After imagining yourself responsible for a castle, would you still take it? Or would you amend your desire: "I want a grand living space, with a sense of remoteness and security, while having all the modern conveniences."

Assume I Have It Now. What Does that Bring Me?

This question catapults you into a richer image of your vision, so you can see its underlying implications more clearly. For example, maybe you wrote down that you want to be the CEO. Why do you want it? What would it allow you to create? "I want it," you might say, "for the sense of freedom." But why do you want the sense of freedom?

The point is not to denigrate your vision thus far—it's fine to want to be CEO—but to expand it. If the sense of freedom is truly important to you, what else can produce it? And if the sense of freedom is important because something else lies under that, how can you understand that deeper motivation more clearly? You might discover you want other forms of freedom, such as what comes from having a healthy figure or physique. And why, in turn, would you want a well-toned body? Or just because . . . you want it for its own sake? All those reasons are valid, if they're your reasons.

Divining all the aspects of the vision takes time. It feels a bit like peeling back the layers of an onion, except that every layer remains valuable. You may never discard your desire to be CEO but keep trying to expand your understanding of what is important to you. At each layer, you ask, once again, "If I could have it, would I take it? If I had it, what would it bring me?"

This dialogue shows how someone handled this part of the exercise:

My goal, right now, is to boost my income.

What would that bring you?

I could buy a house in North Carolina.

And what would that bring you?

For one thing, it would bring me closer to my sister. She lives near Charlotte.

And what would that bring you?

A sense of home and connection.

Did you put down on your list that you wanted to have more of a sense of home and connection?

No, I didn't. I just now realized what is really behind my other desires.

And what would a sense of home and connection bring you?

A sense of satisfaction and fulfillment.

And what would that bring you?

I guess there's nothing else—I just want that. I still do want a closer relationship with my sister. And the house. And, for that matter, the income. But the sense of fulfillment seems to be the source of what I'm striving for.

You may find that many components of your vision lead you to the same three or four primary goals. Each person has his or her own set of primary goals, sometimes buried so deeply that it's not uncommon for people to be brought to tears when they become aware of them. To keep asking the question "What would it bring me?" immerses you in a gently insistent structure that forces you to take the time to see what you deeply want.

Chapter 7 Reflection Questions

Once you've come to think deeply about yourself, your potential, and what you really want to create in your life, it's time to think through what may be holding you back, what obstacles you have encountered in the past that seem to keep you from moving forward with your plans. This brings you to yet another step in your self-assessment. The following exercise is useful for journaling to better understand barriers that keep you from forward movement. Free yourself of any perceived blocks, and you are on your way to identifying the goals that lead toward your vision.

After reading through your entire present situation data about yourself, your values, passions, strengths, and assessment results, ask yourself these key questions:

1. What do you need to *let go of* in order to start your journey toward your vision?
2. What will you certainly want to *hang on to*—taking it with you?
3. What do you see that clearly *needs to be different*?
4. Who will be part of your *support team* on your journey ahead?

Growing Your Leader Self: Seeking Support

In our culture, we pride ourselves on being "doers." We multitask, packing our schedules from dawn to dusk. Our entire families are busy and in a hurry. Virtually everything we do seems to be done in a rush, yet everything we do also becomes a matter of habit. Smoking, drinking coffee, and watching TV are all examples of habitual behavior. We wish we were doing other things that are more enjoyable, healthier, and wiser, such as exercising, playing with the kids, reading, writing a novel, perfecting Tai Chi, singing, dancing, being in nature, learning to play the violin, or simply spending time in reflection.

The cost of this hectic lifestyle is not obvious, but it is great. Most of us do not keep a journal or make time for serious reflection. Why is this important? What is the cost to us of not doing so?

The cost can be tallied by looking at the stress in our personal lives, and that stress is additive in the workplace. It subtracts from both our individual and collective effectiveness. Stress creates dis-ease in organizations, groups, and families and takes its toll on individual lives, often as a major contributor to heart attacks, cancer, and other debilitating diseases.

Most leaders and potential leaders today get so caught up in the hurried pace that they lose sight of the value and benefits that come from stopping and taking time for reflection. At a recent leadership development seminar, when participants were asked to reflect on the course content, there were many who rolled their eyes, as though in disgust. Why would a call for introspection prompt this reaction? Is it because we are all so tied up in the hurried pace that we don't have time or don't want to take time to stop? Or is it because we don't know the benefits to be gained by taking time for introspective reflection?

Most of us are very aware that reflection is a powerful learning tool. It opens doorways to transformative learning. First, it empowers us to challenge our own limiting assumptions and/or socially constructed norms. If we take the time to ask such simple questions as "What is it that I assume?" and "Why do I hold that assumption as truth?" we have the potential to identify our constraining beliefs, to entertain alternatives, and to shift our perspectives. This shift in perspective, followed by a resulting change in behavior, is indicative of transformation. Think of the value in asking leaders to reflect critically on their present leadership practices. By doing so, they are primed for growth and change.

A second value is that reflection is essential to integrating multiple perspectives into our own. In today's environments, we are encouraged to seek feedback from multiple stakeholders in order to improve our performance. So it follows that, when we take the time to understand other people's perceptions of our actions, why they have them, and how those perceptions empower or constrain us, we can grow significantly from the experience. Without reflection, we would be hard-pressed to develop an effective action plan to capitalize on successes.

Many benefits result from taking time for reflection, including reducing stress levels, being able to engage more meaningfully with others, taking on a more responsible inquiry, and appreciating uncertainty. Most people describe leadership as a shared human process, which is essentially an engagement with life and a lifelong commitment to human fulfillment. It is "taking responsibility for ourselves in concert with others, creating a global community worthy of the best that we human beings have to offer" (Bolman and Deal 1995).

The notion of "reflection in action" is the ability to think about what you are doing while you are doing it. It requires that you focus on ideas, make connections, and reach conclusions. Furthermore, it demands that you recognize complexity and multiple perspectives, acknowledging that elusive and messy endeavors are not easily managed, and yet it includes the involvement of many actors. Reflection just might be the pivotal way you learn. It is a way of making learning conscious, and it forces you to be more intentional. It gets to the heart of the matter and helps you make meaning of the past, so that you can more clearly plan for the future. The following are how some of our emerging technical leaders learned from reflection practices:

> Matt Jones said, "I was confused early on about this notion of reflection . . . what was I supposed to be doing, anyway? It was helpful in class discussions to hear what other students were doing—like journal writing on their questions, or spending time in deep conversation with friends and/or associates who are struggling with some of the same thoughts or ideas. I have been able to begin to notice what some of my thoughts are (self-talk) when I am in the midst of having chaotic feelings or a confrontation with a colleague. I have come to see how helpful it is to take time to think through some of those fleeting thoughts or questions that surface in the moments I give myself."

> Barbara Young shared her learning experience of reflection in action. She finds that it most often takes the form of a dialogue with herself. She finds it helps to gain perspective by often thinking back on what she has done, considering how well she did and how she might do it differently the next time. She also does this dialogue on conversations she has had with her direct reports, or peers—thinking through what worked well, what could have been improved, and how she will approach the next steps. She has

found these debriefs with herself very helpful and insightful: "I sometimes
didn't realize how much wisdom I really have when I give myself the gift
of time to truly think deeply." She feels that all of this leads to greater self-
awareness and improved leadership actions.

There was a time in our history when delaying gratification was a virtue.
People saved, they put in hard work to achieve their goals, and they held off taking rewards until later. Our forebearers viewed this behavior as a sign of maturity. Then the pendulum swung.

Today, too often, our society prefers immediate gratification. Witness the
extensive consumer debt, spending personal money we don't have through credit
cards, then suffering when jobs are cut. Is this good, or not? Is it a sign of the
loss of maturity? Is it the abdication of responsibility? Does it come from our
fears that tragedies can happen and we don't want to miss anything?

Whereas the short term is most visible to us, the long term is the big issue.
Will our companies survive? Will our economy be robust? What is our personal
responsibility, and ability, to deal with this in a sustainable, responsible manner?

Educated technical leaders, whether engineers or not, have the responsibility
stated in the Obligation of the Engineer to "be ethical in our dealings, conserve
nature's resources of energy and materials, and serve the public good." You are
responsible for the stewardship and sustainability of our resources. If for no other
reason, this is why technical people bear a huge responsibility to lead. You have
the skills and knowledge to solve the enormous issues facing us in this century,
from environmental and energy issues to health care delivery and clean water.

Almost all of those interviewed are clear examples of people who have
delayed gratification, simply because they have chosen to take the route of fur-
thering their learning, so that their life plan includes more education, more exposure to new ideas and personal growth, regardless of their expected outcomes of
this pursuit. They have clearly chosen to postpone experiences that might have
been more relaxing and fun, so that their education could take precedence. Many
married folks shared examples of wanting to complete their graduate work
before starting their families, and even those who had started families wanted to
complete their degrees before the children were entering school age. Those from
other countries were seeking to get a broad and deep education to help them in
their intended leadership work in their countries of origin.

Persevere with Patience

Betty Jarrett talks about her abilities to stay focused, persevere with patience,
and know that it is in relationships that she finds her success. She has built a
solid reputation, people see her as credible, and she is willing to wait for the
next promotion, trusting from her experience, that it will come. She has seen
that she has succeeded by doing the best possible job at the moment and the
rest will come. She is patient yet perseverant.

Brad Rosen is another example of someone who has been tenacious in his journey. He first recognized his leadership potential as a student, participating as an actor in high school plays. He feels leadership is about choice, self-discovery, personal development, and recognition. He feels everyone has certain innate abilities and weaknesses but recognizes that he has spent his entire life as a voracious learner, reading everything he can find about leadership in order to discern his own leadership style and understanding. He is comfortable with spending time in discovery, balancing personal satisfaction with fulfilling the needs of his present job.

Closely related to delayed gratification are patience and persistence. From the time of Machiavelli and his writings in *The Prince,* we have come to know that trying to change the order of things is difficult. Many people are uncomfortable with change and therefore resist it. Partly because of habit, and mostly because change leads to the unknown, it is difficult to change the way we do things, and even more difficult to change the beliefs that drive our behavior.

Leadership is all about change, so, if you are to lead, you must understand how to help people overcome their discomfort with change. This takes time and patience. Leaders need to raise the issues, expect some push-back, and not expect immediate positive response. Leaders also must understand that messages need to be repeated many times, so that people come to understand and accept before they can eventually change. Many have understood this for decades. This takes persistence. Change is nothing new. What is new is the fact that the current pace and complexity of change are outstripping many leaders' abilities to reinvent themselves and/or their organizations. They are often blindsided by unanticipated events, too often wrestling with complex issues, attempting to create certainty. They have little time to respond to one change before the next wave hits them. Constant change is the new norm, and anxiety its companion. So what are you to do if you want to be an effective leader of change?

John Young, former CEO of Hewlett-Packard, believes that we cannot forecast with any certainty, and that spending time planning for every possible contingency is nonproductive. Young talks of "just-in-time worrying" as an approach. He didn't waste time worrying about things that might never have happened. However, when situations did arise, he and his team put immediate and intensive energy into finding solutions.

Today's leader of change is challenged with becoming a master of emotional capability. That involves facing the unknown with courage and confidence, inspiring and challenging people to do their best, while mobilizing human energy. Learning how to create and manage "just the right level of anxiety" is the new essential tool for leading the emotional side of change (Rosen and Berger 2002). This requires reframing your own view of change and uncertainty and your beliefs about yourself, understanding how to manage anxiety, and having a perspective that reflects realistic optimism, constructive impatience, and confident humility.

During their development process, our technical leaders were challenged to begin to see themselves as masters of leading change—particularly understanding the emotional side of change. Clearly, growing as a leader requires an ongoing focus on growing yourself, since leadership development is self-development. Being intentional about your growth is the most important thing you can do for yourself. Leadership is an art and the instrument is the self. The mastery of the art of leadership comes with the mastery of the self.

This process begins with being very clear about your values, your beliefs, and the principles that will guide you in your decisions. Beyond this critical base is a serious and intentional quest for learning how to learn—both taking in and letting go. This requires self-observation—learning what works for you, learning from others, learning from your teammates, learning from superiors and subordinates, learning from your optimal experiences as well as your failures—learning in every dimension available to you. Only through this learning process do you come to know yourself, to know what you need to unlearn. Finally, you learn you have the capacity to reinvent yourself over and over again.

Pursuit of Continual Reinvention

Our leaders shared their experiences of becoming avid and lifelong learners, in pursuit of continual reinvention:

> Ham Wilson said, "For 18 months, I have used my work as a laboratory for my leadership development, tracking my progress through performance objectives and personal development goals. This process, perhaps not coincidentally, correlates with Otto Scharmer's transformative Theory U. First, I had broken from my previous skepticism, opening myself to honest observation. Next, I reflected on and accepted what I felt to be the emerging truth about my leadership capabilities. Finally, I began acting on my intentions, using my leadership activities at work as opportunities to prototype my actions and responses, receiving feedback before my next iterative assignment. My challenge moving forward is to discipline myself to continue this process, iteratively practicing the steps of observation, reflection, and action. I feel that I am in a unique position to be able to achieve this goal of continual assessment and growth, based on my vision for myself in the future: Every five years I will continue to reinvent and transform myself professionally by continuously planning for the next leg of my journey and adapting to face the new challenges of this continual improvement spiral—one ultimately directed toward a more holistic understanding of the world and my place in it. Over time, this iterative prototyping process will define the trajectory of my leadership development. With each iteration, I will be forced to translate my skills in a new context. I will face new challenges. I will experience abysmal failures and learn incredible lessons. Over time, I expect to see increases in my leadership capacity and my ability to reflect with each reinvention of this learning cycle, eventually tracing the iterative path and overall trajectory of my development."

Carol Jacobs spoke to one of her most central learnings: "In my learning process, I have discovered that I need to let go of some of the misguided beliefs I had about myself as a woman in a man's world. I felt I needed to try to be like the rest of the guys in my group, denying myself of my real truth of who I was. This was a significantly new idea for me to change the way I saw myself as an actor in my world and begin to let my real self show through."

Juan Martinez talked about opening up to his extended family, sharing how important "character work" is for him. He described it as an ongoing building of credibility in being the same person in every environment—home, work, family, and social settings. He said it was important to him to model respect, authenticity, courage, and faith. His model for being a man of character is Abraham Lincoln and he was deeply moved by Lincoln's biography and the testimony that others attributed to Lincoln's character.

Often, others see potential in us before we recognize it ourselves. Who are the people who believed in you before you did?

For me, it began in college when I was an intern in an appliance company laboratory. The lab manager, Frank Sorrentino, gave me responsibilities in failure analysis and product design and stood up for me and a decision I had made and audaciously telexed (long before e-mail) to the manufacturing VP, who then wanted me fired. It continued with Charlie Ring and Lew Coronis at BMC who gave me my first supervisory and executive opportunities, respectively. Jon Swanson at CPI hired me, then recommended I fill his director position when he moved to project management. It continued with former Honeywell vice presidents Clint Larson and Arnie Weimerskirch, and former 3M vice president John Povolny at the University of St. Thomas. Clint devoted extensive time and thought in mentoring me through critical times in forming a School of Engineering, Arnie provided unfailing support in developing strategy, and John has been a mentor for a quarter century and has provided a model for leading an organization with a focus on people.

—Author RJB

As we found during our interviews, some were not told why they were seen as leaders, finding much later that others recognized their leadership potential long before they did themselves. They had an inkling that they had some characteristic that gave others confidence in them. Eventually, they were able to figure it out for themselves. Think how much more they could have intentionally developed these characteristics had they known what they were. Are you telling your employees what special leadership skills they have? Are you helping them recognize and develop those skills?

You need to observe how you are being viewed and, when you get new responsibilities and promotions, ask yourself why. You also need to learn to ask those who make these decisions. Think about your own circumstance. Do you know why you have been given more responsibility? Carol Jacobs does; she asked for it:

> Several leaders, such as Carol, shared their willingness to speak up and ask for new responsibilities, new challenges, and new roles that give them more visibility and opportunities to learn. Very few held back, waiting for someone to notice them and invite them into new opportunities. Rather, it was story after story of people recognizing when the opportunity was there, a new idea was needed, or some creative genius was being sought out. That's the time for putting your hand up. Are you one of those?

There is a wonderful video by Michael Porter of Harvard on the acquisition of Skil Corporation by Emerson Electric. The video shows how MBA students address the issues of the case. It is presented in three sections: first, a lecture on the situation at Skil; second, a section on how the MBA students would handle the acquisition; and, finally, an interview with the Emerson person in charge on how the company did address the issues.

We have used this video in leadership sessions. After playing the first section, we stop the tape and ask the class what they would do. The group identifies the issues and how they would handle them. Then we play the second section, showing MBA students' suggestions. They are quite different from our group's responses. Then we play the final section. The Emerson executive explains his approach: it is almost identical to what our group suggests. After doing this exercise several times since, with the same results, we have concluded that these working technical professionals have all the knowledge and critical thinking skills needed to make tough decisions in a complex industrial environment.

Realizing Your Potential

These technical leaders did not recognize this about themselves. This was a wake-up call to alert them to their potential. Now they just need the tools to recognize the leadership potential within, and to develop the courage to exercise it.

> Twenty-five years ago, when I was a sales manager, we were selling a technical knowledge service. One component of that service was a network of experts who were available by phone. When a client scientist, engineer, or marketing person had a question outside his or her area of expertise, that person could simply pick up the phone and call to get an answer. Managers and early adopters understood the power of this, but
>
> *(continued on next page)*

(continued from page 81)

many technical professionals resisted. Why? It was viewed as cheating. They had been taught they had to figure out everything on their own. Asking someone for an answer was cheating. And, if they did call an expert, they wanted to know everything first, so they didn't look "dumb." Of course, the more confident people found this to be a great asset, and they were recognized for using their time and company resources wisely. They were the leaders.

—Author RJB

Fortunately today, teamwork is not only accepted but also encouraged and even demanded. The notion of individuals knowing everything, or having to learn it themselves, smacks of reinventing the wheel. As one IBM executive said, it's bad enough to reinvent the wheel, but some reinvent the square wheel. Today there is no time to reinvent anything. Learning from colleagues in local teams and around the world is commonplace. Today it's not cheating; it's teamwork. The world is available; there are experts for everything, and they like to talk about what they know. As a leader, you need to use all of the resources available to you. The alumni we've interviewed know that there is power in collective thinking and that other people and the team as a whole can make better decisions than the leader alone.

Dan Jansen talks about the importance of the team, and the role of the leader to create "enablers," whatever will help the team succeed. He explains what needs to be done and why it needs to be done, then gives some ideas. Then he asks, "Can you help me accomplish this?" He creates a culture and an atmosphere for success, noting that it is the leader's job to recognize all those who make contributions and credit success to the team. He worries that far too much emphasis is placed on management and too little on leadership, saying that corporations are a social gathering and it is the people that are important. Understanding this is important to the satisfaction of work, and there is an opportunity for young leaders to change that system.

Wade Dennison believes that leaders need three main characteristics. First is to be respectful of other people, no matter their position or rank. Second is to have an innate skill of looking at systems, seeing how all the pieces interact. And third is an orientation of looking for opportunities and not rushing to judgment. This last characteristic helps attract people who see that their ideas will be heard.

Dick Bastion gets excited about the achievement of others and seeing the team get kudos for their work. He knows the team will come up with the best and right solutions, and he taps into their creativity constantly, challenging them to come up with multiple solutions. He figures the team knows what the problems are and is in the best position to come up with solutions. He is proud that all are working together for the good of the company. Success depends on what they do as a group, not on individual performance.

Brad Rosen said that building relationships that make a difference and getting satisfaction from seeing others succeed and watching them grow are important to him. He considers this viewpoint a matter of maturity, which is a slow process. It takes humility not needing to be the focus of attention.

Corrine Anderson initially found that letting people take responsibility was a challenge for her and she continues to work on it. She now likes to delegate, believing that her team members come up with better ways of doing things than she can alone. She also notes that, as people succeed, they get more passionate about their work.

Keith Kutler is in a major leadership role in his manufacturing company as one of four key executives charged with keeping the business focused. A major challenge is shaping the culture, building loyalty, keeping good people, living out of authenticity and being visible in the organization. He continues getting and giving feedback, listening to the people issues, being sensitive enough to ask for input from people in production while seeking their ideas and including them in decisions. He feels that keeping employees aligned with the organization's goals requires letting people get engaged with major business decisions.

Gaining Confidence

Gaining confidence in one-self is difficult for most of us, even if the signs that others have confidence in us are abundant. Consider how management has treated you. Have you received greater responsibilities? Do you still have your job when others have been laid off? Have you been selected to be a team member on a cutting-edge project? Watch how others respond to you. Do colleagues come to you for advice? Do teammates listen attentively when you speak? These are all signs that others have confidence in you.

If you are not getting recognition, but others have said to you, "You have good ideas; you should speak up more," it means they see something in you but want to see more. You need to step up and make yourself visible. This will help you gain confidence in yourself, as well as contribute to the team and the organization.

One of the most important things you can do for yourself in your continual learning process is to connect with others, gaining their support and encouragement. In fact, without those supporters, you are likely to slip back into complacency. If you do not have a mentor, or coach, seek one out to work with in shaping your leading and learning plan. You will find that having another person invested in your success is very valuable.

Those with mentors know the value they bring in helping to build confidence. Sometimes these experienced people volunteer to be mentors, but more often you need to ask for their help. You can identify people inside your organization, or outside, who you believe have wisdom and knowledge that would be helpful to you. Experience shows that you may have several mentors over your career with different skills and experience, depending on your needs at the time.

Your Personal Board of Directors

Another valuable approach is to build your own personal board of directors—a group of people with diverse backgrounds and experience to assist you in your leadership journey. Your mentor should be part of this board. Choose board members who will be honest and accurate in their feedback and guidance. This board may include your spouse or significant other, a family member who is invested in your success, peers you admire and value, your supervisor—any close friend or associate you respect and know to have wise insights.

It is not necessary to meet with your board as a group, but be clear about why you've asked each person to serve. What gifts and talents does each have to offer? Why do you want those people to help you, and how can they do that? You may want some people to provide you feedback, especially if they are close enough to witness you in your work. You may want to share your leadership goals and plan with these members, helping them see the significant role they might play. That could mean holding you accountable to your goals and timetables, providing guidance on how to get certain things done, or showing you how to handle a crisis situation or difficult conversation. There are endless reasons for you to consider asking others for guidance; people who are invested in your success will be honored to be part of your board. They want to help you and can be counted on as valuable partners on your journey.

Who are some of the people you would like to have on your board of directors? What will you say to invite and engage them in your journey? What skills, attitudes, and knowledge do you feel they could offer you? Make sure to share with them why you value their opinions and insights. Let them know how often you would like to meet with them, making sure to keep your word about regular meetings and engagements, so that you get the best out of your team. Our leaders shared their experiences of using a personal board of directors:

> Cal Archer commented on how valuable his board of directors has been over the past five years. He continues to use them for guidance and advice. Several are family members—his daughter in college, his brother doing graduate work, his Aunt Em (who has been a mentor of his for several years), and a daughter who is teaching. He sees that all of his board members are sincere in their insights and candid with their feedback. He knows how to tap into their guidance regularly and keenly values their investments in him.

> Ken David shared his experience: "I have found that my 'Sounding Board' has been such a helpful suggestion. I am pleased that I have been able to keep the group to a manageable size—my mentor, my father, my spouse, and two remarkable directors in my organization who have been so helpful in my career and seem invested in seeing me succeed. I try to meet with each of them at least once a month and keep them posted regularly on my progress, seeking their feedback and asking them occasionally for specific inputs on some of my projects."

> Patsy Hall feels her board of directors has been helpful in responding to her needs—she uses them as a group of mentors, asking for their advice and guidance whenever she feels she could use another opinion. She says they have been very helpful in opening doors for her, inviting her into new challenging projects, and ensuring she gets visibility in the organization. "Mostly, I know I can count on their wisdom, their faith in me, and they give me straight-shooting feedback and assists."

Setting up your board of directors is a great way to take a risk for your own future. Taking a risk usually means you are innovating, going where you've not been before, and pursuing something that is bound to teach you. If you are to be a change agent, you must try new things. Doing the same old things will just get the same old results. So how do you become less risk averse? The answer lies in assessing what the risks really are. Generally, when you have confidence, have a plan in place for yourself, and know where you are headed, the notion of taking risks becomes commonplace. Taking risks is the way you have decided to take charge of your life and make things happen.

Chapter 8 Reflection Questions

1. In what ways are you being intentional in how you learn, grow, and stretch yourself to step into being the leader you want to be?
2. What kind of support could a personal board of directors provide you? Who might be some candidates?
3. What do you do to be reflective or introspective? What works best for you to maintain the balance between reflection and action?
4. What are some of the lessons you've learned about leading change? How have these had an impact on your approach?

PART 3

Making a Difference

Something has to be done, and it's just pathetic that it is us that have to do it.

—Jerry Garcia

Almost everyone we have ever met has wanted to make a difference with his or her life. Who doesn't? We want things to be better for ourselves, our families, our communities, our organizations, our society, and our world. Each of us may have a different approach and a different focus, but each wants to play a part. We all know there is an endless supply of issues that need to be addressed, so there is no shortage of material to work with. As we discuss in the final chapter of this section, technical professionals have a particularly special and important role to play because of their technical expertise, professional responsibility, and internal passion to make a difference.

In this section, we look first at "being the change we want to see in the world" and then allowing that to stimulate us into committing to doing our part through action learning, building our roadmap ahead (while ensuring that we stay balanced with reflection and action) and building strong and healthy relationships. ●

Be the Change You Want to See

Mahatma Gandhi said, "You must be the change you want to see in the world." You cannot expect someone else to take charge. You need to accept the responsibility to make it happen. It will take initiative, courage, and extra effort. You can do it, and here's your chance.

Start with a reflection on the key question: "What is the change you want to see in the world?" There are so many problems in our society that it is difficult to identify the most important ones. Don't wait for someone else. Notice what is going on, search for solutions, and, most important, take action. What is yours to do?

The hour is striking so close above me,

so clear and sharp,

that all my senses ring with it.

I feel it now; there's a power in me

to grasp and give shape to my world.

I know that nothing has ever been real

without my beholding it.

All becoming has needed me.

My looking ripens things

and they come toward me, to meet and be met.

—Rainer Maria Rilke

Imagine what will happen if you step up and take charge. Dare to become what you want to be, and decide what is yours to do. When you began your journey into technical education, what was your motivation? Did you hope to solve some big world problem? Did you expect to win a Nobel Prize? Was there some experience you had as a child in school, in 4H, or with parents that stimulated your interest in technical subjects? After your technical education, when you began your professional career, did you have a long-term goal or seek out certain kinds of jobs or companies? As you have matured and gained experience, have you developed other interests? In those experiences are the passion and purpose that give meaning to your life. Do you know consciously what that passion is?

Some of the leaders we interviewed had strong and well-developed passions early on, and they developed life plans to work toward their goals. Most did not; their passions were revealed as they gained more experience. Whatever the case for you, now is the time to reflect on this question.

Marsha Salter wants challenges and has sought them out. She has felt powerful as a result of asking for new challenges. Her passions are energy, conservation, new power source development, and medical products. She likes her present company because it fits well with her passions and values; the company is challenging, values aggressive research, and invests in progressive equipment. She feels it is a place where she can learn about new materials, new processes, and new tools. She has found in her experience that most larger companies offer these things.

Corrine Anderson, in giving advice to younger people launching their leadership journeys, says, "Understand what your passion is and how to use it in getting what you want. Figure out what gets you fired up, ask for opportunities to lead, and invite others to join you in what is exciting for you. You will witness them finding their passions as well."

At the heart of your emerging confidence and leadership, inside your perspective on the world, are your beliefs and passions. When you find them and identify them, you will have discovered your true self. As we mentioned earlier, one way to identify your passion is to think of a subject you always care about—something you can lose track of time studying or doing.

"We are called to the place where our deep joy meets the world's deep hunger."

—Frederick Buechner (1993)

Matching your skills with your passions will lead to fulfilling your dreams. What events of the past puzzled you, especially what others have said about you or your potential? Stop here. Put down the book. Find a quiet place without interruptions. Think for 15 minutes about this and write it down. You will have a good start at waking up and finding what's inside.

Welcome back. Now, consider the following stories. They may help you understand how others have identified their passions and looked for ways to make a difference with their lives.

Monica Rogers is passionate about her learning and wants to "move the needle" for herself and others in their own thought processes and technologies. She has deliberately and intentionally organized a broad network of people who are thought of as leaders in their fields, and they constantly share new knowledge, new technologies, and new trends in their respective businesses. She strives to create sustainable new procedures and processes and provide leading edge consulting in risk management. She values digging deep into herself to become more aware of her strengths and capabilities, to know where her limitations are and then strengthens herself through constant learning.

Parallel to passion is commitment—the conscious ability to put your own needs second in order to serve a purpose, an idea, or another person first. Obvious examples include ordinary people risking danger to help others, but the circumstances and your actions don't need to be dramatic. Here are a few other stories of commitment:

Ray Adams feels that a very important principle in his work is to truly know that the central role of a leader is to be in service to his or her people. He says, "One needs to treat people at all levels with mutual respect for the jobs they have to do, listen to them to understand, be fair, and follow up on your commitments to them, helping them to understand the big picture view."

Joe Monahan, a project manager at a high-tech company, is constantly asking himself what he can do to lead his teams more effectively. He feels strongly that people management is the key to high productivity and great results. He sets high expectations for his teams, works to establish trust, and spends several hours of coaching time each month with each of his people to help them improve their skills and their personal development. He challenges them to do things that will make them proud in hindsight. He believes in his people.

Becoming an Authentic Leader

Leadership is about serving others. It's about serving a cause that is important to you and your organization—focusing on a goal, not on competitors or obstacles.

In True North (2007), Bill George says, "To become authentic leaders, we must discard the myth that leadership means having legions of supporters following our direction as we ascend to the pinnacles of power. Only then can we realize that authentic leadership is about empowering others on their journey."

A transformational shift happens when we are truly making a difference. It is a shift from "I" to "we." It is the only way we can unleash the power of all members of the organization, enrolling them in "making the difference" as well. If subordinates are merely following our lead, their efforts are limited because it is our vision, not theirs. An effective leader enrolls followers in creating a shared vision and direction and thereby builds the "we" throughout the organization—all motivated by and aligned with one vision.

Leadership is about building an organization that supports the end goal and knowing why it is important. As former National Academy of Engineering president, William Wulf, noted in his 2006 speech at the University of St. Thomas, creativity is the process of combining two ideas that have not been connected before. Since each person has a different perspective and experiences, the more diverse the group that is brought together, the more diverse the ideas and the more creative the group. Diversity in any aspect is of value, whether it is in culture, age, geographic origin, gender, education, or work experience. Innovation is the process of allowing creativity to flourish. There will be many differences of opinion within such a diverse group. The job of the leader is to keep the focus on the goal.

In Chapter 2 we examined the Value Creation Model. To encourage a diverse group of people to be innovative requires an environment and a culture that is open and fosters creativity. It is the leader's job to create this culture. The same culture-building effort can be within small workgroups as well. Big things come by example and can lead the whole organization to change. Recall from Chapter 2 the example of Bobby Bridges at a truck assembly plant and his influence on the entire international organization.

If we are going to be the change we want to see, we need to have done our work on recognizing our own self-inflicted limiting beliefs, changing habits that do not serve us or others, and coming to recognize our sense of personal power.

Do you understand and take ownership of your personal power? What does it mean to have personal power, and how do you recognize it? Personal power is more important and more effective than position power, even if you have a position with many reports. It is the trust and confidence others have in you and your leadership that get them to follow you. There are endless stories of people with position power that have not been good or effective leaders.

The Hagberg Model of Personal Power

Janet O. Hagberg (2002) defines personal power as a combination of external power (a capacity to act) and internal power (a capacity to reflect). Her Personal Power Model distinguishes **six stages of personal power and leadership** in organizations, as shown in Table 9.1 and Figure 9.1.

In this model, leaders at each stage develop followers who are or want to be like them. Each person's quality is more important in determining leadership than position or status. A true leader, according to Hagberg, has experienced a crisis of integrity and reached Stage 4 or higher. However, such people often do

TABLE 9.1 Hagberg model of personal power

	Stage	Characteristics	Leads By	Manages By	Motivated By	Needs from Manager
Externally oriented		Power primarily sought and obtained from outside the person, from titles, positions, or other symbols or status				
	1. Powerlessness	Secure and dependent, low in self-esteem, uniformed, helpless but not hopeless	Domination, force	Muscling, force	Fear	Support, direction
	2. Power by Association	Apprentice, learning the culture, dependent on supervisor/leader; new self-awareness	Sticking to the rules	Maneuvering, catching up	Learning	Safety, freedom to explore
	3. Power by Achievement	Mature ego, realistic and competitive, expert, ambitious	Charisma, personal persuasion	Monitoring results	Visible signs of success	Feedback, challenge, questions
		Power primarily sought and obtained from the inner journey of the person				
	4. Power by Reflection	Reflective, confused, competent in collaboration, strong, comfortable with personal style, skilled at mentoring, showing true leadership	Modeling integrity, generating trust	Mentoring or coaching process	Inner exploration	Time, space
	The Wall	Moving beyond your intellect, letting go of control, embracing your shadow, going to your core, finding intimacy with your higher power, glimpsing wisdom				
Internally oriented	**5. Power by Purpose**	Self-accepting, courageous, calm, conscience of the organization, humble, practical mystic, elusive qualities, generous in overpowering others, confident of life calling	Empowering others, service to others (servant-leadership)	Acting as catalyst	Living their calling	Protection
	6. Power by Wisdom	Integrating shadow, unafraid of death, powerless, quiet in service, conscience in the community/world, compassion for the world	Wisdom, a way of being	Musing	Self-sacrifice	Nothing

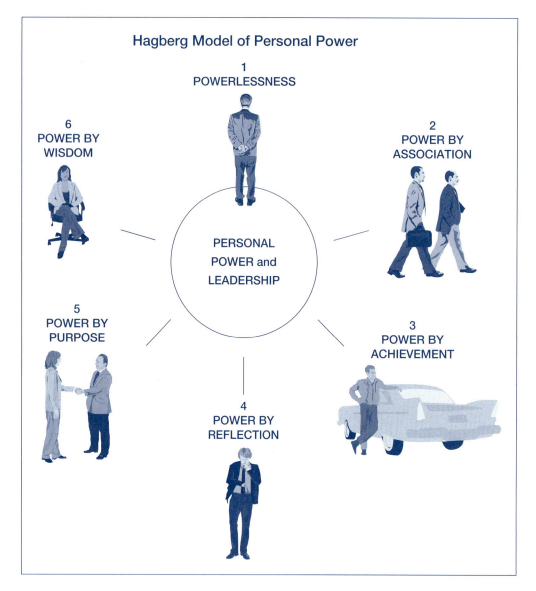

FIGURE 9.1 Personal power and leadership (Hagberg 2002).

Source: Reprinted with permission of the publisher. From Real Power: Stages of Personal Power in Organizations, by Janet Hagberg. Published by Sheffield Publishing.

not pursue positions of power for their own sake and may even shy away from them. The following assumptions underlie Hagberg's model:

- The stages of personal power are arranged in a developmental order.
- Each stage is different from all the others.

- People can be in different stages of power in different areas of their lives, at different times, and with different people. However, each person has a representative home stage.
- One can move through the home stages only in order from 1 to 6.
- Power is described and manifested differently at each stage.
- Each stage contains positive and negative dimensions, as well as developmental struggles.
- Women are more likely to identify with certain stages, men with other stages.
- People do not necessarily proceed to new stages merely with age or experience, although both are factors.
- The most externally and organizationally oriented power stages (1–3) show a marked contrast to the internally oriented power stages (4–6).
- The development and then release of the ego are central tasks inherent within this model. Cultural rituals are necessary in order to do that successfully.

Carol Jacobs explains how, early in her career, she was placed in a "high-growth" management position and was in a supervisory roundtable, but she didn't feel she would be effective at managing people. She loved being a technical person. She says the technical side has a ladder, too, but it is more challenging. She recognized that the management side had benefits of income and title, but status didn't matter to her. She was more interested in job satisfaction and would turn down a technical director job. She knows herself, and she loves "specializing in being a generalist." Her influence on outcomes is high. She believes you don't need to be a manager to be effective. You also don't need to know everything or be an expert; you just need to know whom to go to. She loves being the "go to" person and, in that way, exercises her personal power.

In a highly matrixed team environment, project managers and directors seldom have many direct reports, yet they are responsible for getting projects finished. With so little position power, they must rely on personal power and influence skills. Their ability to build trust and communicate is key to their success.

Mac Casey is a project manager at a Fortune 50 company, where he has no clear authority over the many people within his influence. He manages suppliers, original equipment manufacturers (OEMs) in the automotive industry, and project teams through influence. He is a capable communicator with good interpersonal skills and is able to model doing the right things, inviting others' participation, and making his teams feel pride in their achievements. He clearly knows how to use his personal power to make things happen and to help others feel they are contributing in meaningful ways.

When Dan Jansen was asked how he got things done, he noted that people liked and respected him for including them in decisions and keeping them informed. He said, "When I walk down the halls, people acknowledge me immediately and always say 'Hi.' That is in contrast to the executives around here . . . employees totally ignore them." This is a perfect example of embodying personal power.

Most work is done in teams today, many of them cross-functional with multiple disciplines, discourse groups, and social styles. As a team leader, you are responsible for keeping the team on task while drawing on their skills, abilities, and creativity to achieve a goal on schedule.

Think of leadership as service. As a leader, your role is to help the team, even while you need to use influence to succeed. You provide resources and remove barriers. Your most important job is to develop relationships with team members and build trust. A close second is to do the same with stakeholders. Treating everyone as a customer is a helpful perspective.

Are you thinking creatively about the future you envision? Some people prefer to work within current structures; others constantly look for unconventional ways to pursue their goals. Managers can take either approach, but leaders must make changes. Even improving the performance of an existing process requires doing things differently.

Take Bea Ellison's advice: do the unusual and surprise people with the results. It may not seem like much to you, but, as one of our colleagues put it, "Changing the color of paint on the walls is radical around here."

Orrin Matthews is a leader at a major medical device company, where he is becoming known as the knowledge expert in Program Project Management (PM). He is co-chair of the Project Management Forum, whose strategy is to influence education and knowledge about PM. In both this environment and his internal environment, he is constantly challenging others by demonstrating creative thinking. He sees the furious pace of work and change all around him, and he loves helping others find new ways to improve their productivity, invent new processes that reduce development cycle times, and save everyone time. He is high energy, high paced himself, and eager to bring his creativity to the center of the issues.

When you are thinking creatively, what kinds of ideas come forth? Most technical professionals understand that at the heart of any new idea is a failure. Engineers know that almost anything can be done better. Any unsolved problem is an opportunity for trying something different. When you complete a project, list what went well and what didn't. In the second list are the seeds of new ideas. What could be done differently to avoid problems in the future?

Dan Jansen was working in a major defense company when he was operating as a manufacturing engineer on two major military aircraft programs. When he saw that lead times and yields on the A1 program were not good, he gathered some colleagues. Using learning tools from his graduate program, he proposed changing the way the upcoming A2 program should be run. He recommended "point of use stores" near production, pull vs. push methods, disposition of nonconforming product on the spot, using modern manufacturing thinking and methods. He didn't ask permission; it was just the right thing to do. He sold the approach to those not familiar with these methods; because he had thought them through well, his colleagues saw the merits and joined him. Dramatic improvements followed. It took 180 days to get the first six products finished; after one year, they were producing one per day.

As you identify new ideas in which you are interested, ask yourself several questions: What is the idea? What problem does it solve? Why is it important? How would the solution add value? As you talk with others, if someone says, "That can't be done," or "We tried that 10 years ago and it didn't work," don't give up. Has something changed that would make it work now, or could it be modified to work? As you pursue the idea, determine what additional skills you need from others to make it work, and identify who can help you. Mentors can be very useful here.

As an exercise, think of yourself as a bicycle. The drive train and rear wheel provide the power; the front wheel steers you in the direction you want to go. Your rear wheel is your technical skill—carefully machined, well maintained, and in excellent working condition (see Figure 9.2). Without steering, though,

FIGURE 9.2 Bicycle model of leadership.

this power won't take you anywhere. You need a front wheel to control your direction. This means right-brain elements, such as reflection, leadership skills, communication skills, courage, initiative, creativity, and innovation. These skills make your bicycle complete and will keep you on the right path. During the development of your technical skills, the skills needed for your front wheel may have been neglected. You can fix that.

What change do you want to see? What is your vision for the future? What is the direction of your long-term plan? How can you build on your experiences and the relationships you've established? Once you answer these questions, you will be prepared for creating your roadmap.

Throughout history, all change has come from the power of an idea, usually through one person who is passionate about the idea, becomes its champion, and persistently pursues it. We often think of major changes, but, in fact, these are made up of many smaller changes. No matter the scope or size of the change, it still takes a leader. For your passion, that leader is you.

The needs of the world are at your doorstep—issues such as energy, the environment, clean water, cost-effective health care, causes of terrorism—the list goes on. What can you do about these issues? Maybe you think you can't do much, or that someone else will take care of them. But you know they are too important to be left to chance.

These are very personal issues. For example, think about energy. How does it affect you, your family, and your community? What can you do about the issue at a local level? Where do your skills come in, and which skills are relevant and important? If you can do something in your family or in the community, what else can you do in your job?

The fundamental issue that needs leadership at every level is nothing less than sustainability. The National Academy of Engineering published a book in 2005 titled *Rising Above the Gathering Storm*. It spoke of these major issues and how we must address them. The title implies we need to act now, before the storm arrives. If it meant for us to wait, it would be titled *Rising Into the Treacherous Storm.*

At the heart of the solution for these issues is technology. No one is more knowledgeable about technology than technical professionals. So who should lead the charge to their solution? You.

A main theme of this book is "waking up to leadership." Waking up means recognizing the signals around us calling for leadership. What unfilled needs can you see? In each unfilled need is a call to wake up, a call for you to take a leadership role.

Sometimes these wake-up calls come as sudden inspirations. In just one moment, you recognize a need *and* your ability to do something about it. They may also be gradual realizations. Have you ever spent a lot of time thinking about something, then wakened at night with a new revelation about it? This is your moment to be the change you want to see in the world.

Chapter 9 Reflection Questions

1. As you think about the change you want to see in the world, how does this affect what you imagine is yours to do?
2. Think about how you exercise personal power. What are some examples of how you influence others?
3. List some examples of creative thinking at your workplace. How is it encouraged? When are you moved to try new approaches?
4. What "next great idea" are you excited about? How will you pursue it?
5. What is the first small step you can take toward making a difference?

Action Learning

We've discussed the critical importance of ongoing learning as a major piece of developing your leadership capacity. In this chapter, we look at some methods that can keep your learning active, as well as intentionally making a difference in how you contribute.

Action learning is learning from experience drawn from real problems in your organization, your community, or your society. The stories from emerging leaders in Chapter 9 are examples of action learning.

Imagine you are a research physicist. Just when years of painstaking effort seem about to bear fruit, you come up against an intractable problem. You are surrounded by other research scientists, all of them experts—but none in your field. Does this stop you from seeking their help? Not if you want to solve the problem. You share it with them and find out whether they can help you.

In the 1930s, a young man named Reg Revans found himself working with just such a high-powered group at Cambridge University. When they were faced with difficult research problems, they would sit down together and ask one another lots of questions. No one person was considered more important than any other, and they all had contributions to make, even when they were not experts in a particular field. In this way, they teased out workable solutions to their own and each other's problems.

Revans was so taken by this technique that, when we went to work at the Coal Board, he introduced it there. When pit managers had problems, he encouraged them to meet together in small groups on site and ask one another questions about what they see in order to find their own solutions, rather than bring in "experts" to solve their problems for them. The technique proved so successful that the managers wrote their own handbook on how to run a coal mine.

This is how action learning was born; but it was some years before Reg Revans, now Professor Revans, presented the cogent theory that is now the cornerstone of many leadership development programs. Today it is used widely by organizations of all sizes, industries, and locations. Both private and public sector organizations have found amazing success using action learning.

The methodology follows a specific equation: $L = P + Q$. The L stands for *learning;* the P is for *programmed knowledge;* and the Q is for a focus on *questions.* By focusing on the right questions rather than the right answers, action learning emphasizes *what you do not know* rather than what you know. Action learning tackles problems for which there are no easy answers but may result in a number of possible solutions. For learning to take place, you need to do more than just experience solving your problems effectively; you need to reflect on the experience in order to identify exactly what you have learned, internalize the lessons, and devise action plans, so that you can take effective action in the future in new situations.

Action learning is a specific methodology that focuses on action and reflection, and it suits most emerging leaders well by helping them solve problems in their workplaces and learn from the experience. These leaders then review and interpret their experiences to identify what they have learned. It follows the familiar experiential learning cycle:

Action → Reflection → Planning → Action

The process has four basic features:

1.   Problems to work on
2.   An action team of "colleagues in adversity"
3.   A team leader/coach to facilitate the learning process
4.   A sponsor who owns the problem, issue, or task

The sponsor may be the leader's supervisor, but it's more important for the sponsor to want and need results.

This methodology was taught and used by graduate students at the University of St. Thomas. They were required to identify an action learning project in their own organizations that could become opportunities for them to become more visible, to tackle significant problems, and to lead an action team to results. It was familiar to some who had participated in "skunk works" teams, other action teams, or quality team efforts. For others, it was more challenging. Here are some of their experiences:

> Lincoln Canning's action learning project at a small engineering firm was to define safety guidelines for guarding nip equipment. When his firm acquired another companies product line, he noticed an ongoing problem with designing all guarding to Category Three standards set by OSHA, which led to excess costs. His goal was to define a safety methodology that was applicable to the firm's equipment. His team included himself as leader, the assembly supervisor, a representative from the safety committee for his firm, and a manager from coating. The sponsor was his immediate manager. He was successful in accomplishing the project goals within the targeted timelines. His primary learning in the process was how important communication skills are in a cross-functional team and how effective the deep questioning process was. Because of his success, he has quickly gone on to another action learning project—to update the firm's mechanical design review process, which was seriously in need of revision.

Bill Bonson's action learning project at a major medical device company involved 22 members of a cross-functional team working together to implement an improved catheter-tipping process. The project involved a $1.3 million proposal for better equipment. After six months of work and many meetings, Bill says he learned valuable lessons about navigating organizational politics, persevering in the face of setbacks, and listening to those you trust. He said, "All members of the team learned from the experience, particularly how to treat one another with respect." Three months after the project was completed, Bill was promoted to a project manager position.

Barbara Johnson's action learning project focused on developing and implementing a peer-mentoring program that began as a department-wide process and ended up organizationwide at a major medical device company. Although this project was beyond the scope of her engineering responsibilities, she felt as a member of the Employee Satisfaction Committee that it was an important issue with the potential to benefit many others—particularly women in technical jobs. She actively led a cross-functional team to design and implement a process that has been recognized as highly useful. She is an active peer-mentor herself and a model for others. The project helped her gain confidence in her abilities to make things happen, using her keen sense of how to bring people together to shape a process that has high value. She felt she grew in her stature as a leader and was suddenly elevated within her own department.

Sam Sorenson's project at a major manufacturer of recreational vehicles was to make navigation of the culture in his company less of a mystery for new employees in his department. In short, he wanted to enable new employees to "learn the ropes" more quickly. He felt this would benefit all new employees and his own department. He involved a group of engineers who had been hired in recent years, along with several managers of new employees. His end result included identifying a navigator to serve as a mentor/coach for each new employee. Orientation training was designed, and the program has been replicated in other departments. His learning relates to his attitude: "When you see a problem, step up and take action." He has been affirmed in his belief that anything is possible.

Taking Risks and Seeking Support

Each of these projects represented courage: taking risks, seeking support, and pursuing a vision of what is possible. When asked how each developed the courage to step up, they answered in different ways. Some were just bold by nature.

Others were quiet and needed to develop confidence. Some were put in situations where they effectively had no choice. There is no one answer. It depends where you are on your leadership journey, the environment you are in, the strengths you have to build on, and the new experiences that call for your leadership.

Some choose to start their action learning projects in their own workgroup, some in volunteer organizations. Whatever your choice, gather your courage and get going. You probably will find you have much to gain.

Although change may be risky, lack of change is even riskier. Successful organizations realize that change is constant. Leaders need to be aware of what is happening internally and externally and be prepared to move quickly. Their organizations need to be nimble, fast-paced, and ready for rapid change. Those who cannot respond in time may lose the chance to catch up later.

During interviews, we asked alumni what recommendations they would make to their younger selves based on their experiences. Relevant to you as well, their advice is explored in greater detail in Chapter 14; here is a summary to help guide your development plan:

- Get to know your passion, your own individual desire.
- Believe in yourself and your ideas.
- You don't need to know everything, just how to learn.
- Understanding your weaknesses helps you understand others.
- Engage in experiences that will change your mindset.
- Put aside all-or-nothing thinking; become empathetic and listen actively.
- Apply what you learn about yourself.
- Become a T-shaped leader, with a broad generalist overview.
- Understand the strengths of, and best roles for, each member of your team.
- Become good at networking.
- Become aware of different social styles.
- Find a nonprofit or civic organization that needs your leadership.
- Figure out what you want to lead, what gets you fired up.
- Find a mentor.
- Don't replace sound judgment with "cold" analysis.
- Skate with your head up; look for opportunities.
- Be respectful of all others, no matter their position or rank.
- Do the right thing; act from basic ethics.
- Speak up for others.
- Apply your skills in a different department.

We authors have come to know a number of entrepreneurs, each of whom started with a good idea. They all say that, when they began developing their ideas, they were unaware of all the problems they would face, yet each overcame barriers and became successful. As you begin your journey of leadership, you cannot know what lies ahead. But you can be confident in one thing: you will learn and grow as you pursue your goal.

The learning experience will be more intense than anything you've done during your formal education, since it will be driven by urgency. There is no time to lose in tackling new problems, particularly in a competitive environment where someone else has surely seen the same problem and is undoubtedly pursuing solutions. Consider the just-in-time worrying approach used by John Young as a model for action learning.

Like all entrepreneurs, you are the sole proprietor of your own career. You need to take charge of that journey, seeking out all the learning and growing you can to succeed. No one else will . . . or can.

> As a young engineer, I thought someone would come to me and ask me to take on a challenge. I would sit quietly at my desk, watching action going on around me, but no one ever asked me to help. It never happened. I liked action and got bored easily. It was boredom more than anything else that spurred me to get involved. Whatever the driver, you cannot wait for someone to ask you to help. You need to take the initiative and ask for responsibility, or "response ability." You must put yourself in an uncomfortable position and become visible.
>
> —Author RJB

> Corrine Anderson recalls that she did not recognize the need to speak about her career interests as a young engineer; however, the company expected people to talk about their career aspirations. She took the opportunity when she was promoted to director. She made herself visible and is now in an environment where she can be seen regularly by decision makers. She said, "One can lead at any level and have influence . . . You can interject yourself into a leadership role at higher levels, but you need to develop confidence in yourself." She recommends that everyone speak out and ask for roles of responsibility. Sitting back and waiting for it to happen does not work.

Our growth and learning as leaders is seldom smooth or linear. For many, it is a series of spurts, often driven by seemingly random factors. You might be available to lead a project when your company gets a new contract; or your boss might be promoted, leaving you to replace her; or you may be chosen to serve on a select team to develop a new product at record pace. You can plot every detail of your career path, but it often shifts in new directions. As one of our colleagues says, *no amount of careful planning can beat sheer dumb luck.* These unique situations are opportunities to step into leadership.

Think about pivotal moments in your career—times when you were offered more opportunity and responsibility or times when you saw a situation that needed to be addressed and just took charge. What factors caused those pivotal moments, and what did they mean to you? Do you know why you were selected? What behaviors were you demonstrating? Why did others have confidence in you? Were you told, or did you figure it out? Or do you still not know?

The example in Chapter 8 of the decision-making capability of graduate students in the Emerson Electric/Skil Saw acquisition case is profound. It demonstrates the point that experienced, practicing technical professionals who consider themselves ordinary have extraordinary abilities and judgment. Most did not recognize they had this ability and, if they did, would not even admit it to themselves. Part of this is modesty.

More importantly, they never tested themselves. They did not step up to leadership and take on big problems. They did not pursue opportunities to prove their value, recognize their competence, and build their confidence. They short-changed themselves. The stories in this book are from technical people who did take action. They discovered more knowledge and skill within themselves than they once thought they had. Mac Casey even described his leadership experience as fulfilling and euphoric.

Taking a stand means speaking up in situations where you haven't done so before. However, there are times when you need to be heard. As others have said, speaking up can give you energy and confidence. The first time is the most difficult, but no one else can speak for you.

Tim Torino, a project manager, found he was perceived as being indirect with people. This led to others mistrusting him and keeping their distance. "I was playing off what I thought others would value in me and trying to project that image," he said. This confused others. He found he needed to start speaking his truth more directly, sharing his opinions, and listening more closely to others' opinions. This honesty led to feedback, both positive and negative, about his style and his values. He had to become more vulnerable as well, admitting mistakes and faults more readily. After three years of deliberate and intentional work on how he was affecting others, he found a place that felt very comfortable and authentic, not only to himself but also to those around him.

Gene Paul said, "I have developed courage in the face of a significant amount of image management that goes on in the office. I am not afraid to challenge the status quo, when everyone around me is just saying yes. I make it known how I feel about changes that impact others, mainly the less fortunate around me, i.e., younger managers and staff."

Learning from Mistakes

Despite what some say, no one ever succeeds alone. Everyone needs help from others. Even if we could figure out everything ourselves, it would be a poor use of time and effort. If what you need is already available, why go through the costly and time-consuming process of designing and making it yourself?

The same concept applies to learning. If you have access to people who already have experience you don't, why not go to them for advice? This is the idea behind having a mentor. And because one person as a mentor also cannot know everything about everything, a group of mentors with different backgrounds and experiences is even more powerful.

We all have, and often forget, another source of advice and knowledge as we move from job to job throughout our careers: our networks of friends and acquaintances. When we face new situations and challenges, chances are that our network can help. Carol Jacobs, as noted in Chapter 7, has used this practice effectively throughout her career.

No matter how long you've been in any career, you can't know everything. Through your formal education and experience, you recognize what you know; through experience, you also know what you don't know. You might never have thought about what you don't know that you don't know, but it's there—and it's infinite.

Years ago, people thought that, when they finished schooling, they were through with learning. Later, they realized that was just the beginning. You need to keep moving forward, looking for more knowledge and awareness. As with physical fitness, you can never just stay in one place.

> In addition to being an engineer and leader, Ray Adams is also a weight-lifter. He believes that in leading, like weightlifting, there is no such thing as maintaining. Either you're getting better or getting worse.

Think back to the 6th-grade girls' summer camp experience in Chapter 7. Although the camp is designed to introduce girls to science, engineering, and manufacturing, they identified three other things they liked about the camp compared to school. One, they could work in teams. Two, they could fail and then learn to recover. And three, they knew *why* they were learning specific subjects.

The point about failure was highlighted during the first week of the first year of camp. Girls built and flew radio-controlled, gas-powered airplanes. Many crashed. The next day, all 40 girls did a failure analysis, identifying the nature of the failures and the possible causes. It was probably the best learning experience we could have provided, as well as the most effective at showing what technical professionals do in their work.

The question is, how do you fail without disaster? There are times in a project where failure is costly, like the first flight of a full-size aircraft. But research, as noted by a neurology researcher, is all about failure. When you are trying new things, looking for new discoveries, failures are commonplace. The key is whether you choose to learn from them and change your approach.

The advantage of having failed is that it teaches you to recognize similar situations in the future. Think about your own failures. What have you learned from them? The story of Dan Jansen and the wind shear project in Chapter 7 is an excellent example of learning from failure.

Does failing make you a bad person? Many executives of successful companies encourage innovation and, when failure occurs, ask two things: first, that the person experiencing the failure notify others immediately and, second, that the person doesn't have the same failure again. That's it. In a culture where people are rewarded for innovation and honesty about failure, individuals develop confidence and bounce back, puting extra effort into their work the next time.

> Gene Paul took on an action learning project involving the design and implementation of a software system that was supposed to cost $2 to $3 million. The estimate then ran to about $12 million and was going up about $1 million a week. The project suffered on several fronts, and in the end it was pulled by senior management. "I could have done better during this project to trust my instincts 100% of the time," Gene said. "As it turned out, my feelings or premonitions about processes and people were mostly correct." He also learned to be more prudent in his comments and personal feelings about people. The project and some of the learnings attached to it gave Gene new insights into the environment of a large retailer and in many ways altered how he approaches projects today. He trusts himself more completely and knows how to stop the process when he thinks people are not committed to the changes or other problems surface. "Being open to the lessons of a major project are significant for all involved—there is much for all to learn."

We can be our own worst enemies. We are self-critical, never satisfied with our accomplishments and our work. We are cautious and avoid taking chances. We are poor judges of our own work and worth. So why do others see more in us than we do ourselves?

How can you see yourself as others see you? Watch their behavior; how are they responding to you? Use inference to figure out why. Ask for critiques from others, and accept them positively. The people who appear to be your severest critics may be your best allies.

As Warren Bennis describes in *Geeks and Geezers* (2002), "crucibles" are a metaphor for circumstances that transform a leader. More often than not, crucibles are times of failure with potential for learning. They are defining moments that force choices, sharpen awareness, and teach us about the kind of people we really are. Some individuals can be destroyed by these experiences; others see them as opportunities. According to Bennis, "leaders create meaning out of events and relationships that devastate non-leaders." When they encounter failure, leaders do not see themselves as helpless; they look for what is useful, and how they might respond with appropriate actions.

Every great leader has experienced these defining situations. What's more important is reflecting on the outcomes, then using the new insights in the future. Our interviews are filled with stories of crucible moments that became opportunities for learning.

Chapter 10 Reflection Questions

1. What opportunities have you had to take on action learning projects?
2. Can you think of a project to propose today that would give you opportunities to be more visible in your leadership practice and solve company problems at the same time?
3. When have you taken a stand for what you believe is ethical and right? What pushed you to take that stand, and what happened as a result? What did you learn?
4. What have been some of the greatest learning experiences in your life? How do these correlate with your experience of drawing your lifeline and noting the peaks and valleys?
5. Think of a crucible experience in your life as a leader. What happened? What did this learning experience teach you? How have you changed your approach as a result?

CHAPTER 11

Drawing Your Roadmap

Even though meaningful and important changes often happen by chance, you may not notice them unless you pay attention. With intentional awareness, you can integrate these changes in deliberate ways. Part of creating and sustaining effective leadership is to recognize, manage, and direct your process of learning and change.

People who consciously direct their development are better prepared to make choices that help them become more effective and satisfied in life. Richard Boyatzis and colleagues (2008) have developed an Intentional Change Model, which helps people engage in transformation and embrace it. The model includes several key elements (see Figure 11.1):

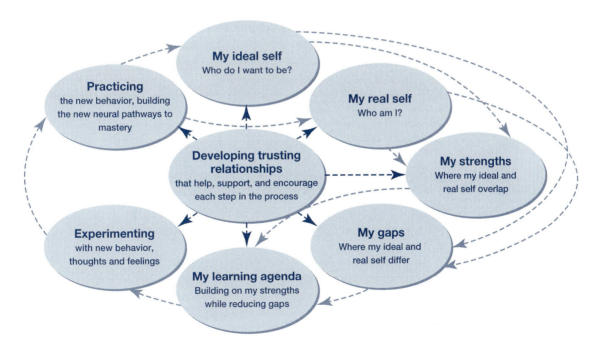

FIGURE 11.1 Intentional change model.

1. The **ideal self**—what you want out of life and the person you want to be—leading to your personal vision
2. The **real self**—how you act and are seen by others; the comparison of the real self to the ideal self results in identification of your strengths and weaknesses—leading to your personal balance sheet
3. Your **learning agenda**—building on your strengths and moving you closer to your vision while possibly working on a weakness or two
4. **Experimenting**—practicing new habits or leadership behaviors in order to reinforce or affirm your strengths
5. **Developing trusting relationships**—maintaining close, personal connections that enable you to move through this cycle toward continual renewal

Change begins for you as you discover the leader you want to be—that is your ideal self. You know you have reached it when you suddenly feel passionate about the possibilities your life holds. To take this step, identify your dream for yourself, your life, and your work.

You also need to confront your real self and your present situation. This includes getting feedback from others about how they see you, then comparing that information to the way you see yourself. You need to make yourself vulnerable, avoid defensiveness, and prepare to move out of your comfort zone into new practices and experiments. Those things take courage. Talking with close friends or your mentors can help, particularly if you agree to be honest and supportive and to honor each other's confidence.

Once you identify the gaps between your ideal self and your real self, you can draft an agenda for creating your future. The agenda should focus on development itself—on learning first and outcomes second. An orientation to learning reinforces your capabilities and expectations of growth. To keep your momentum, choose only a few things to work on. Select four or five goals that point you in the direction of your ideal self and your leadership vision. Here are some examples of a few leaders' learning plans:

Bea Ellison identified six major milestones she wanted to pursue in her five-year plan for her learning and leading agenda:

- Choose a new employee to coach and mentor; continue annually
- Participate on a project team where I will be responsible for engineering processes that are outside my current realm, so that my breadth of experiences expands
- Take on a team leadership role for a new technology project in the next year
- Take on a challenge to interact more with the business and marketing teams to learn and influence product development strategies in the next two years
- Assume a leadership position at a senior level in my division where I have responsibility for decisions of strategic importance to guide new technology and product development
- Start a family within the next five years

Keith Kutler's major goals for his leading and learning plan were

- Learn to be a better communicator
- Learn to take chances/have more faith
- Become more understanding and empathetic
- Let go of control and rely on others

Graduate students were invited to use either the Intentional Change Model or the template shown in Figure 11.2 to begin putting together their learning agendas and subsequent leadership development plans.

Your Personal Strategic Plan

As you can see in Figure 11.2, your future state and present states are at opposite ends of the plan. What remains is how you move from one to the other and who will support you on the way. The roadmap serves as a template for your

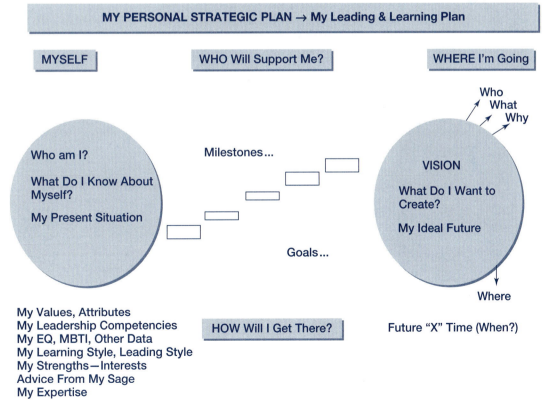

FIGURE 11.2 Millam model for developing your leading and learning plan.

thinking. How will you get from your present state to your future state? What timetable are you willing to commit to? Will it be a plan for three years ahead, five years, or simply one year to get started?

You need to be the judge of what works for you in putting together your plan. Prepare for experimentation and practice, and get feedback from colleagues as you begin your plan. Use your board of directors as wise counselors and coaches to help hold you accountable for what you want to make happen.

> As any change must begin somewhere, it is the single individual who will experience it and carry it through. The change must indeed begin with an individual; it might be any one of us. Nobody can afford to look around and wait for somebody else to do what he is loath to do himself.
>
> —Carl Jung

Karl Conner's leading and learning plan for two years ahead was focused on growing as an authentic leader. He chose five areas on which to focus:

- Understand and live my purpose
- Practice solid values
- Lead from my heart
- Establish connected and enduring relationships
- Demonstrate self-discipline

Sam Antigliota's five-year plan was to continue on his journey of personal mastery:

- Become more of an expert in my field
- Continue to attend conferences and seminars that are relevant to my job and career path
- Gain a new perspective on my job and societal contribution by volunteering to work with a doctor who is involved in the area of interventional cardiology
- Practice all leadership competencies on a daily basis
- Stay networked with respected leaders that influence me and make me a better, more competent leader and person

Each of these plans included a clearly identified and described vision statement, a set of guiding principles, identified strengths, timetables for goal accomplishment, a plan for how each would use a board of directors, and specific actions for overcoming obstacles and reaching goals. A template for the development plan is included in the appendix.

Once your agenda and plan point you in the right direction, you have to put them into action. Go beyond your comfort zone to practice new attitudes and behaviors until they become automatic responses. Every early win sparks new motivation, which in turn brings new energy and commitment. To practice and experiment, find and use opportunities for learning and change; this is easier in conditions that feel safe. After each period of experimentation in a safe setting, practice the new behaviors in the settings where you plan to use them—at home or at work.

Relationships are essential parts of our environment, and they are central to sustaining any personal transformation. The most crucial relationships are those with your personal board of directors. They include the people most invested in your success. They can serve as critical guides and sounding posts, providing feedback as you need it.

Our culture, our reference groups, and our relationships mediate and moderate our sense of who we are and who we want to be. We develop and elaborate our ideal selves from these contexts, as well as label and interpret our real selves from them. In this sense, our relationships are mediators, moderators, interpreters, sources of feedback, and sources of support, and they give us permission to change and learn.

In our interviews, leaders spoke again and again about the value of their support groups, their mentors, and the feedback or guidance those people were able to provide. This affirms several research studies that suggest having a circle of support surrounding you as you attempt to grow and develop will increase the likelihood of your achievement of goals and will stimulate you to attempt even more audacious goals for the future.

Chapter 11 Reflection Questions

1. What are your intentions for ensuring your continuous growth and development?
2. How can you assure yourself that you are ready to take action and not overwhelmed with too much data?
3. What strengths do you intend to leverage?
4. How will you ensure these leveraged strengths will reduce any areas of weakness?
5. What are the first few steps you are willing and ready to take?
6. What stands in your way?
7. How will you monitor your change process?
8. Who will be helping you assess your progress?
9. How will you use your supporters wisely?

Relationships Are Key

Previous chapters described processes for you to learn about yourself, create a vision for what you want, and develop a roadmap to take you there. One more key is building relationships. Whatever your goal, you will need help.

You will need to convince others that your goal is important, that you need their support to achieve it, and that they will benefit when you pursue and achieve your goals. You need to know how to manage relationships well, not only for your own advantage but also for the benefit of the whole—the team, the organization, the society, or the world at large.

Emotionally intelligent leaders can manage both themselves and their relationships with others. They are able to move people with compelling visions or shared missions. Leaders embody what they ask of others and inspire them to follow. They are able influencers, finding just the right appeal for a given listener to create buy-in from key people and build a network of support.

Leaders with influence are persuasive and engaging when they address a group. They know how to have a positive impact on others, generating an atmosphere of respect, helpfulness, and cooperation. They draw others into active commitment to collective efforts and build spirit and identity. They spend time forging and cementing close relationships beyond work obligations.

Part of this ability includes communication skill. Emerging leaders need to have their ideas seen and heard, to report findings from research, to share information, and so on.

There are many ways to communicate—in writing through memos, e-mails, reports, research papers, and other documents; orally through speeches, presentations, and discussions; and visually through presentations, graphs, and photos. You choose the method depending on the nature of the material, the audience, and the amount of time available.

You may think of communication in terms of giving information to others; just as important is how you receive information, listening actively and using open-ended questions. As several leaders have said, their main job is influencing others. Using your emotional intelligence to build relationships is not random; it is a deliberate process you can learn. The better you understand how to build relationships, the more effective you will be as a leader.

Relationship building has two primary aspects. One relates to social styles for you and your audience; the other involves understanding group dynamics and using effective communication.

In their book *Social Style/Management Style* (1984), Robert and Dorothy Grover Bolton established a framework of social styles based on simple yet powerful concepts: a person in each of the four major social styles, which we will discuss in detail later in this chapter, thinks differently and places importance on different criteria for developing trust, for personal and professional motivation, for what they must do, and for what they fear. If you understand these differences, you can develop communications that speak effectively to each social style.

Larry Wilson (1987, 1994) and Wilson Learning build on the social style concepts developed by the Boltons. This process is well documented in the works of others, including William Murray (1993), Peter Block (2008), Peter Senge and colleagues (2004), social science theorists, human relations and organizational development practitioners, and trainers.

Some communication concepts are not immediately obvious, but they are crucial to your effectiveness. Fortunately, these concepts appeal to the technically trained: you need to measure only two variables, assertiveness and responsiveness, and from these you can infer other relevant information—and the process can be replicated and followed. The big picture is more complex, but this approach is based on work by the Boltons in *Social Style/Management Style* and taught by organizations such as Wilson Learning and Eagle Learning—and it works.

Your life is a complex matrix of relationships with friends, family, teammates, members of professional societies, religious organizations, and more, which take time to build and need care to maintain. The process of building and maintaining relationships seems random until you understand how to manage it.

Do you get along better with some people than others? Why is there a difference? Is it you or them? When you have good interactions with people, and feel you are being heard, what characterizes that experience? What about when you have a bad interaction with people, and feel you are not being heard? These questions get to the source of the influencing process. You have been influencing others all your life and probably just aren't aware how.

To effectively influence others, you must be able to read the dynamics of a situation and manage any tension between yourself and others, whether they are customers, employees, or colleagues. All human interactions involve two kinds of tension. One is *relationship tension,* the everyday tension all humans experience. The other is *task tension,* a productive force that drives us to achieve our goals. Effective influence reduces relationship tension and increases task tension.

People love to buy things, and they value group membership. However, sometimes it's difficult to "sell" an idea, even to your own community. You may have been on the other side of a situation like this, where you were unmoved by someone else's idea. We all have barriers that create resistance. By helping others break habits and accept change, you can be more effective at accomplishing your goals, adding value to yourself and your organization, and feeling that you make a difference.

To build productive relationships, you must establish trust. People who trust you will listen to you. Similarly, when you show interest in others and listen to their ideas, you will be able to identify what goals and needs you have in common. By explaining how your ideas benefit others, you get them moving in your direction. To summarize, establish trust, identify needs, propose help, and create a positive sense of urgency.

Social Styles

This is where the concepts of social style become important. Nearly 50 years of research (Bolton 1984, Wilson 1987 and Murray 1993) show that people operate with four distinct ways of interaction, or social styles: Amiable, Analytical, Driver, and Expressive:

- **Amiable style** people are people-oriented and care more about close relationships than results or influence. They usually appear warm, friendly, and cooperative. Amiables tend to move slowly and deliberately, minimizing risk and often using personal opinions to arrive at decisions. Belonging to a group is a primary need, and Amiables work to gain acceptance. They typically seek common ground, preferring to achieve objectives through understanding and mutual respect rather than force and authority. When managed by force, Amiables appear to cooperate initially but will likely lack commitment to the objectives and may later resist implementation.
- **Analytical style** people value facts above all and may appear uncommunicative, cool, and independent. They have a strong time discipline coupled with a slow pace to action. They value accuracy, competency, and logic over opinions, often avoiding risk in favor of cautious, deliberate decisions. Analyticals are usually cooperative, providing they have some freedom to organize their own efforts. Power often raises suspicion in Analyticals, but, if they come to see it as necessary for achieving goals and objectives, they may seek power themselves. In relationships, Analyticals are initially more careful and reserved, but once trust is earned they can become dedicated and loyal.
- **Driver style** people want to know the estimated outcome of each option. They are willing to accept risks but want to move quickly and have the final say. In relationships, they may appear uncommunicative, independent, and competitive. Drivers tend to focus on efficiency or productivity rather than devoting time and attention to casual relationships. They seldom see a need to share personal motives or feelings. Drivers are results-oriented, tending to initiate action and give clear direction. They seek control over their environment.
- **Expressive style** people are motivated by recognition, approval, and prestige. They tend to appear communicative and approachable, often sharing their feelings and thoughts. They move quickly, continually excited about the next big idea, but they often don't commit to specific plans or see things through to completion. Expressives enjoy taking risks. When making decisions, they tend to place more stock in the opinions of

prominent or successful people than in logic or research. Though they consider relationships important, their competitive nature leads Expressives to seek quieter friends who are supportive of their dreams and ideas, often making relationships shallow or short-lived.

Establishing Trust

Starting a relationship requires someone—you—to take the first step. You need to take a psychological risk in trusting first, assuming you can trust the other person. This is often self-fulfilling: if you trust others, they will likely trust you.

In building trust, remember that words are only a small part of communication; you also send messages through body language, voice, empathy, and credibility. This is why face-to-face conversation is effective, and why misunderstandings are more likely in written communications.

Communication experts claim that only 38% of our communication is realized through words, whereas the majority comes through pace, intensity, and nonverbal cues such as facial expressions, gestures, eye contact, posture, and tone of voice. What you say matters; how you say it matters even more.

The way you listen, look, move, and react tells the other person whether you care and how well you're listening. These nonverbal signals can produce a sense of interest, trust, and desire for connection—or disinterest, distrust, and confusion.

Practice body language behavior. Observe responses when you smile or maintain eye contact, two powerful actions. Watch skilled actors. When they are not speaking, they are still sending messages and having an impact on others. Many people send confusing or negative nonverbal signals without knowing it. To improve the quality of your relationships, you need to

- Read other people accurately, including their emotions and unspoken messages
- Create trust and transparency by sending nonverbal signals that match your words
- Respond with nonverbal cues that show others you notice, understand, and care

Understanding and using nonverbal communication will help you connect with others, express what you mean, navigate challenging situations, and build better relationships at home and work.

In addition to the words you use, your tone, pitch, volume, inflection, rhythm, and rate are important. These sounds provide subtle but important clues to your feelings and meaning. Tone of voice, for example, can indicate sarcasm, anger, affection, or confidence. Listen for these elements in the way others speak, and in your own voice:

Intensity. The amount of energy you project is considered your intensity. This has as much to do with what feels good to the other person as what you personally prefer.

Timing and pace. These show your ability to listen well and communicate interest and involvement. Pace is also an indicator of assertiveness.

Understanding. Sounds of acknowledgment, with congruent eye and facial gestures, communicate an emotional connection. More than words, these sounds are the language of interest and compassion.

Tone of voice. This indicates responsiveness. In general, Drivers use a fast monotone; Expressives use a fast sing/song; Amiables use a slow sing/song; and Analyticals use a slow monotone.

Empathy. Sympathy is sharing others' feelings; empathy is the ability to see things from others' perspectives. It is the ability and desire to understand people's moods, temperament, or feelings, to read people accurately and accept them as they are. With an understanding of these aspects of communication, you can establish trust and credibility with various social styles. Here are some examples:

- For Drivers, credibility is based on demonstrated competence: knowing your subject, describing your proposal succinctly, or responding confidently to rapid-fire questions.
- For Analyticals, your credibility depends on your image—how you appear and sound. For example, do you dress appropriately for the situation and do you speak like a professional?
- For Expressives, credibility is linked to commonality. Do you come from the same state? Did you go to the same schools? Do you belong to the same organizations?
- For Amiables, competence is closely tied to positive intent. Instead of pushing your product or idea, do you have the best interests of your client at heart?

Earlier in her career, Ellie Fitzgerald was a project manager who had constant personality problems with the quality manager. Nothing she did seemed to be right. She learned to understand the social style of this manager and, through persistence and active listening, broke down the communication barriers in three months. She discovered that her previous approach was not perceived by the quality manager as demonstrating respect. They are now enjoying a great working relationship as the result of Ellie's observation and perseverance.

Identifying Needs

Once you establish a base of trust and manage relationship tension, you can begin building task tension. Technical professionals often have great ideas and believe they are very important, but the ideas alone may not be enough to get support from others. This is known as "technology push." By listening for the needs of your audience and linking your ideas to these needs, you can improve your chances of success. This is called "market pull."

You can have all kinds of reasons for someone to buy your ideas and products, but they may be irrelevant. People buy for their reasons, not yours. For example, ask several people what cell phone service they use. You will get a variety of answers. When you ask what they like about it, the answers will also vary. It doesn't matter what the cell phone company thinks is most important; it may be something the company doesn't even advertise.

When you are excited about a new idea, remember that others may not be, no matter how good the idea. Determine what they view as important; then put your idea in terms that speak to them.

Recall Ellie Fitzgerald's experience in Chapter 2. She had a situation and an idea for handling it—but her group had divergent opinions on how to respond. She facilitated a meeting, asked questions, and used active listening to get all her team members to share their thoughts. As each team member was heard, the best ideas surfaced and the team agreed on one solution. Ellie achieved her goal.

Most buyers want three things: an adequate solution, a trusted consultant, and value-added service. The adequate solution needs to address the buyer's business motivators (task issues or advantages) and personal motivators (feelings or benefits). These vary with each social style. Advantages, or business motivators, are such things as increased profits or investment; increased quality or quantity; reduced effort; and lower costs. Equally important are the buyer's personal motivators, or benefits, such as respect, acceptance, recognition, and power. Here are the motivators for each social style:

> **Drivers.** Business motivators for Drivers are increasing quality and being the best. Personal motivators are increased influence and power.
> **Analyticals.** Business motivators for Analyticals are increased profit and sound investment. The primary personal motivator is gaining respect.
> **Expressives.** Business motivators for Expressives are reduced effort. The primary personal motivator is gaining recognition.
> **Amiables.** Business motivators for Amiables are lower costs. The primary personal motivator is being accepted.

Because you will need to seek support from a variety of people, you should phrase your proposal in a way that addresses all of these motivators.

To identify needs, ask questions like a doctor trying to find the source of a medical problem before suggesting a course of action. Don't offer a solution before you identify the need. Start by looking for gaps between the current situation and a better one, with questions such as these:

- What is the situation now; what do you have?
- What is the future desired state; what do you want?
- What is motivating your desire for change?
- What are the potential consequences of inaction?
- What barriers must be overcome to get what you want?

A satisfied need or want is not a motivator. For instance, suppose you just finished an expensive dinner and couldn't eat another bite. The chef offers you another just like it for free if you eat it now. Because your want is already satisfied, you have no motivation. Remember this when searching for needs and wants for your idea, product, or service.

Avoid questions that can be answered by yes or no. Open-ended questions allow others to give you more extensive answers that keep the conversation going. Use questions such as "What is your current status relative to schedule?" or "How do you plan to complete this project?"

Using words such as *want, wish, desire, need, think, dream, could, feel, opinion, change,* and *enough,* you can discover wants in a similar way. Ask, "What do you want?" or "What are your dreams?" or "How would you go about getting to your goal?"

Knowing what someone has and wants, you can determine the task or business motivators, as well as the feelings or personal motivators, that drive the desire for a different future state. This leads to questions and answers about the cost of the problem in terms of revenue, schedule, poor quality, and so on. What is it costing to not achieve the desired future state?

Proposing Help

Once you have established trust and identified needs, think critically about whether your idea, product, or service will help. Assuming the wants are not already fulfilled, continue to build task tension. The final stage of diagnosis is to determine what obstacles are in the way of action. Why hasn't anyone taken steps to achieve the future state? What is keeping someone from doing so?

Listen for relevant information, watch body language, and pay attention to nonverbal cues. In this Socratic process, your questions help audience members clarify and articulate their thinking and, in doing so, reveal the issues to themselves. Use active listening responses, such as "I see," "Tell me more," "Give me an example," and "Then what?" Make sure your body language and nonverbal communication also show your interest.

When you understand the reasons, rephrase them in your own words and ask if you have things right. Being able to summarize the situation means you are communicating successfully. Even if you miss a key point, the other person will correct you.

Returning for a moment to the idea that "people buy for their reasons, not yours," you now know what their reasons are. By questioning, you help people determine what they really believe and want. Once they say it, it's their idea. Now is the time to plan and practice your presentation.

Before offering a solution, take some time to look back at the analysis from your active listening process. The idea, product, or service you propose should have three components: a recommendation, an advantage that relates to business or task motivators, and a benefit that addresses personal or feeling motivators.

A **recommendation or feature** is a product, service, set of materials, or proposal for change that satisfies a need. In your recommendation, share *what it is or how it works or functions. Warning*: technical professionals have a tendency to stop there, thinking the implications are obvious. But this is just the beginning. The major work is yet to come. An **advantage** is how the recommendation solves the task or business problem. It tells how the feature provides *increased profits,* *higher quality, greater quantity, or reduced effort.* A **benefit** is how the buyer will *feel* once the product, service or idea is meeting the need or solving the problem. These feelings will be of *power, respect, recognition, or acceptance.*

When you meet again, review the analysis, confirm the needs and wants expressed, and ask if the situation still exists. This refreshes memories and creates opportunities to discover whether anything has changed that would affect your proposal.

If there is still some resistance once you present your proposal, don't be discouraged. It's time to ask more questions. Did your proposal miss the mark? Is another barrier in the way? Because you have formed a trusting relationship, this conversation can be friendly and comfortable. Remember that your goal is to solve a problem in order to meet needs and wants.

Establishing Urgency

After you establish trust, need, and help, decision making can still be a hurdle. This is sometimes known as decision anxiety. The person or group whose support you need may have returned to high relationship tension and low task tension. Your job now is to minimize their feelings of risk.

Show that you are a trusted helper; explain how moving forward with your proposal will help solve problems, how its benefits and advantages will serve your audience's best interest. If delaying a decision will make the situation worse or cost more, say so.

Preparing for the Fight-or-Flight Response

Decisions lead to change, which introduces new risks. And the decision you want someone to make is just one of many things the person may need to handle. Other work, family, and personal matters affect their choices as well. When people are under stress, they respond differently, with fight-or-flight reactions. Prepare for these reactions, and be ready to offer reassurance.

Reactions, like most behaviors, are related to social style. Fight reactions are manifested in an autocratic or attacking response. Flight reactions are manifested in avoiding and acquiescing. As the influencer, avoid putting yourself under stress. You understand what is going on and can help, which is an advantage for you—and for your audience.

When I was a flying student working toward a single engine pilot's license, my instructor would often pose hypothetical problems and ask what I would do. What if the engine failed? What if I got into a spin? What if we hit a bird? The answer was always the same: fly the plane. I remember that lesson every time I get into a stressful situation. I still need to maintain control of myself, even if things around me are failing. I need to fly the plane.

—Author RJB

Building Self-Confidence

For others to have trust and confidence in you as a leader, you need to have trust and confidence in yourself. When you trust your own objectives, knowledge, and judgment, you can "fly the plane" even when things are going wrong.

That literally happened in January 2009, when US Airways flight 1549 lost power shortly after take-off from New York's LaGuardia Airport. Captain Chesley "Sully" Sullenberger landed the Airbus safely in the Hudson River, saving the lives of everyone aboard the plane. Through years of experience and training for emergency situations, Captain Sullenberger had the confidence and courage to deal effectively with a critical situation that could have become a catastrophe. His crew had trust in him, as he did in them.

Developing confidence in yourself means reducing your fears of failure and rejection. Many people avoid taking action, even on things they want to do, because they are afraid to fail.

What have you put off doing that you know would help you in your job? How about in your personal life? Why haven't you done those things? Part of the answer relates to back-up style; another part is irrational worry.

Back-up style is triggered when something unexpected makes us react as if threatened. An activating event, A, which is a surprise, triggers our belief, B, resulting in consequential feeling, C, which causes us to do something, D. Our feelings drive our behavior. When they are positive, we are at our creative best; when negative, we are at our worst. The C's and D's are not the result of the A's, but of the B's. Beliefs that are not in our best interest have words such as *must, ought,* and *should.* Beliefs based on myths are unreal; they cause improper or inadequate actions (Murray 1993).

Irrational worries also inhibit action, even when the worst consequences are not that bad. What if you fail? It will be inconvenient. What if you make a mistake? It will be inconvenient. What if you are rejected? It will be inconvenient. What if you have to face some pain? It will be inconvenient. Not awful, not a tragedy, not the end of the world; just inconvenient.

Look deeply into yourself and shed your myths, so that you can take the actions you know are needed. Begin with confidence in doing what is right, and you will build more confidence in yourself.

Influencing others involves intellectual and emotional skills. It is not easy at first, but it becomes simple when you learn to have confidence and manage your emotions. You can create personal influence by following a well-defined process and measuring two variables. Like most processes, this one takes time to perfect. The time to start is now.

Handling Conflict in Relationships

Conflict is inevitable. How it's handled can bring people together or tear them apart. Poor communication, disagreements, and misunderstandings can create anger and distance or can lead to stronger relationships and a happier future.

Next time you deal with conflict, keep these tips on effective communication skills in mind and you can bring about more positive outcomes:

Stay focused. Sometimes it's tempting to bring up old conflicts when dealing with new ones. Unfortunately, this often clouds the issue and makes finding mutual understanding and a solution to the current issue less likely, and it makes the whole discussion more taxing and confusing. Stay focused on the present, understand one another, and find a solution.

Listen carefully. People often think they're listening when they're really thinking about what to say next when the other person stops talking. Effective communication goes both ways. Although it might be difficult, try really listening to what others are saying. Don't interrupt. Don't get defensive. Actively listen and reflect back what the person is saying, so that he or she knows you've heard. Then you'll understand the person better and he or she will be more willing to listen to you.

Try to see the other person's point of view. In conflict, most of us want to feel heard and understood. We talk a lot about our point of view to get the other person to see things our way. Ironically, if we all do this all the time, there's little focus on the other person's point of view, and nobody feels understood. Try to really see the other side, and then you can better explain yours. If you don't "get it," ask more questions until you do. Others will be more willing to listen if they feel heard.

Respond to criticism with empathy. When someone comes at you with criticism, it's easy to become defensive. Although criticism is hard to hear, and often exaggerated or colored by the other person's emotions, it's important to listen for the other person's issue and respond with empathy. Also, look for what's true in what the person is saying; that can be valuable information for you.

Own what's yours. Personal responsibility is a strength, not a weakness. Effective communication involves admitting when you're wrong. If you share some responsibility in a conflict, look for and admit to what's yours. It diffuses the situation, sets a good example, and shows maturity. It also often inspires the other person to respond in kind, leading you both closer to agreement.

Use "I" messages. Rather than saying such things as "You really messed up here," begin statements with "I" and make them about yourself and your feelings—for example, "I feel frustrated when this happens." It's less accusatory, sparks less defensiveness, and helps the other person understand your point of view rather than feeling attacked.

Look for compromise. Instead of trying to win an argument, look for solutions that meet everybody's needs. Through either compromise or a new solution that gives you both what you want most, this focus is much more effective than one person getting what he or she wants at the other's expense. Healthy communication involves finding a resolution with which both sides can be happy.

Take a time-out. When tempers flare, it can be difficult to continue a discussion without having it become an argument or a fight. If you feel yourself or others getting too angry to be constructive, or using destructive communication patterns, take a break from the discussion until you both cool off. Sometimes good communication means knowing when to take a break.

Don't give up. Although taking a break from the discussion is sometimes a good idea, always go back to it. If you both approach the situation with a constructive attitude, mutual respect, and a willingness to see each other's point of view, you can make progress toward the goal of a resolution to the conflict. Unless it's time to give up on the relationship, don't give up on communication.

Tips to Remember

- Remember that your communication goals should be gaining a mutual understanding and finding a solution that pleases both parties, not winning an argument or being right.
- It's important to remain respectful of other people, even if you don't like their actions.

The most critical piece of building and sustaining relationships is awareness of your impact. You learn this by noticing the effects of your communications, behaviors, and actions. How do others respond? Are they eager to join with you in making a difference in the world, or do they shy away from contributing when there are opportunities to engage? Those who can communicate optimism to the people around them inspire others. By expressing hope, you can build strong and trusting relationships. People will be attracted to join whatever endeavor you propose.

Communicating the Data vs. the Story

The provost of the Texas A&M University, Professor Karan Watson, said at the 2010 American Society for Engineering Education meeting plenary lecture that data are necessary to support conclusions and technical arguments, but they are not enough. To sell an idea, you need a story.

Many people in technical fields believe data *are* a story, to which the conclusion is obvious. This is a mistake that can get in the way of understanding. To engage others, show how your ideas are relevant. Tell why an action or a proposal is important, what problem it solves, and what the implications are if it is not done. You still need data, of course, but make sure you have a story.

Groups of experts in any field develop shorthand communications filled with words and phrases that carry significant meaning for those in the group but not for those outside the group. To communicate effectively with broader audiences, use common words and phrases that are understandable outside your technical sphere. You also need to use the appropriate media.

University of St. Thomas Professor John Abraham has used data from other experts very effectively to present a case for global climate change, as well as to stimulate worldwide interest in the topic. He has gone well beyond the data and told the story better than others. He has catalyzed a rapid response team of experts to respond to erroneous information presented in the media and is changing people's minds—even those adamantly opposed to believing that we are facing global climate change a short time ago. As Suzanne Goldenberg wrote in the *Manchester Guardian,* Abraham and his two colleagues "have come off almost as climate science super heroes, which in a sense they are." (guardian.co.uk, Monday, 22 November 2010)

In communicating your story, start with a clear understanding of the audience and the venue. What method of communication is best—an article, a speech, a presentation, an e-mail message? How about length? Do you have two minutes with an executive and need an elevator speech? Will you be selling your story one on one and need a personalized approach? Will it be a presentation to a large group of people with varied backgrounds where you need to be succinct yet cover a lot of social styles and discourse groups?

A powerful method of storytelling is through appropriate graphics. These speak to all audiences, including technical professionals. Make use of visual communications, such as graphs and photos. Illustrations often communicate complex data better than words or tables. Some of the most highly regarded books on visual communication are written by Edward Tufte (Tufte 1983). He provides excellent examples of storytelling through graphics, including one that shows Napoleon's entire invasion of Russia of 1812 in a single picture (see Figure 12.1). Probably

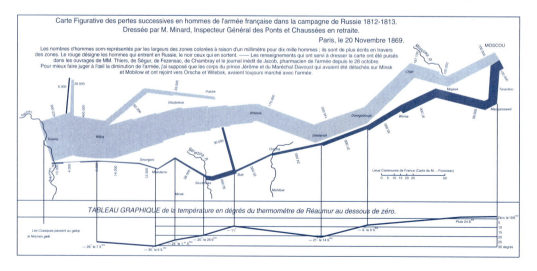

FIGURE 12.1 Example of graphical communications used for complex topics (Tufte 1983).

Source: Reprinted by permission, Edward R. Tufte, The Visual Display of Quantitative Information (Cheshire, Connecticut, Graphics Press LLC, 1983, 2001).

the best statistical graphic ever drawn, the map, by Charles Joseph Minard, portrays the losses suffered by Napoleon's army. Beginning at the Polish-Russian border, the thick band shows the size of the army at each position. The path of Napoleon's retreat from Moscow in the bitterly cold winter is depicted by the dark lower band, which is tied to temperature and time scales.

Using the Active Voice

Many technical professionals learn to write in passive voice. This comes from an academic convention in which the agent is less important than the event being described. However, the active voice conveys greater confidence, credibility, and power. For example, "we performed a test" sounds better than "a test was performed." The active voice takes responsibility, whether it results in credit or blame.

Building strong relationships is an essential requirement for you to achieve professional excellence, to gain personal satisfaction, and to make a difference. No one can do it alone. In order to build relationships, you need to hone your listening skills and develop a range of communications skills, knowing when each is appropriate and how to use them effectively. As we point out, to achieve your goals, all you need to do is help others achieve theirs.

Chapter 12 Reflection Questions

1. How would you describe your communication practices?
2. How do you engage others or share your ideas in order to enroll them in your ideas, plans, and/or vision?
3. What areas of your relationship building do you want to grow and develop further?
4. How will you do this? Be specific in the areas you want to develop.
5. Who are some of the leaders you know who have the ability to draw people into their visions and ideas?
6. What do you notice in their behaviors and actions that are compelling for you and others?
7. What are your gifts and talents in building trusting and strong relationships?
8. How do you exercise your gifts and talents in making a difference?
9. What kind of feedback do people give you about your strengths in relationship building?

PART 4

Why the World Needs You

In this section, we discuss the broad societal need for science, technology, engineering, and mathematics (STEM) professionals to hear and accept the call to address the big issues of this century. These people are uniquely qualified to lead in an increasingly technologically complex world. We describe the ethical obligation a scientist or an engineer commits to, as well as what it means to take this obligation seriously. We look at the meaning of leadership and how it aligns with the needs of our future. We speak to the global demand for innovation and innovators, and we show how scientists and engineers play key roles in leading the process. Finally, we explore the sustainability of leadership practice and outline the way a leader with broadened perspectives can work collaboratively with others to bring about solutions for a world that needs them. ●

The Call to Leadership

During the 20th century, a truly amazing number of technical advances occurred that addressed human needs and altered people's lives forever. These began as ideas and dreams and were brought to reality by pioneers who devoted their lives to improving the human condition. These creative, innovative scientists and engineers changed our individual lives and leveraged their inventions to benefit our entire society.

The National Academy of Engineering has published the 20 engineering achievements of the past century that most transformed our lives (Constable and Sommerville, 2005). This list demonstrates how engineering advancements based on science have affected the way we live. The top 20 are

1. Electrification
2. Automobile
3. Airplane
4. Water supply and distribution
5. Electronics
6. Radio and television
7. Agricultural mechanization
8. Computers
9. Telephony
10. Air conditioning and refrigeration
11. Highways
12. Spacecraft
13. Internet
14. Imaging
15. Household appliances
16. Health technologies
17. Petroleum/petrochemical technologies
18. Lasers and fiber optics
19. Nuclear technologies
20. High performance materials

Despite these advances, many human and technical problems have yet to be solved. Industry and academia have identified 13 major issues requiring science and engineering solutions in the 21st century; many correlate with the National Academy of Engineering (NAE) *Grand Challenges for Engineering* (National Academies NEWS, Feb 15, 2008) listed on the next page.

These issues were developed in a 14-month project. The NAE convened a select, international committee to evaluate ideas on the greatest challenges and opportunities for engineering.

In another initiative, among the key issues for the 21st century identified at the 2006 Engineering Deans Institute are

- Demographic issues: population, education, food, poverty, and disease
- Natural resource management: energy, water, and the environment
- Globalization: democracy
- Infrastructure: transportation, communications
- Security: terrorism and war
- Breakthrough technologies: many areas of biology, nanotechnology, etc.

As noted, the National Academy of Engineering recently published the 14 Grand Challenges for this century, including environmentally friendly power, the capture of carbon dioxide, countermeasures for nitrogen cycle problems, and reverse-engineering of the brain. These are some of the major issues engineers and scientists need to address, as well as the solutions for which they will be responsible. It is in your own strategic interest that you fully understand these issues and your role in managing them. The quality of life on earth will depend on it. Here is the full list:

1. Environmentally friendly power
2. Nuclear fusion
3. Capturing the carbon dioxide
4. Countermeasures for nitrogen cycle problems
5. Quality and quantity of water
6. Reverse-engineering the brain
7. Computerized catalogs of health information
8. Develop new medicines
9. Counter the violence of terrorists
10. Sustaining the aging infrastructures of cities and services
11. Improved methods of instruction and learning
12. Computer-created virtual realities
13. Enhancing exploration
14. Reducing vulnerability to assaults on cyberspace

As Dr. Joe Ling, the former 3M executive responsible for the Pollution Prevention Pays program and member of the National Academy of Engineering has noted,

- Environmental *issues* are *emotional.*
- Environmental *decisions* are *political.*
- Environmental *solutions* are *technical.*

Not only do environmental issues follow this format, but so do all the major issues of the 21st century. Any new initiative meets resistance, simply because change is difficult. Scientists and engineers must become strong leaders to acknowledge the emotional and political aspects of these issues and to move us toward technical solutions to our problems. The future of our world is too important to leave these major decisions to the uninformed and the unwilling. In the long run, these areas represent our social and economic well-being, which means major business opportunities as well. The sooner they are addressed, the better.

This means important roles and potential rewards for new scientists and engineers. But the traditional image of technical professionals is not exciting to many in the public, especially to young students. Given the opportunities in science and engineering to improve the human condition, it is important to change that image.

The National Academy of Engineering released a study titled *Changing the Conversation* (National Academy of Engineering, 2008). This 18-month study concerns the public image of science, technology, engineering, and mathematics (STEM), particularly engineering. This publication demonstrates "right-brain" vs. "left-brain" thinking and provides valuable information on how to get students, parents, and the public to embrace STEM. In focus groups conducted during research for this book, one of the phrases that resonated with parents and students was "Engineering . . . because dreams need doing."

What role should you play, as a technical professional, in stimulating interest in STEM pursuits among young students? How can you promote technological literacy for the general public?

Scientists and engineers—in fact, all technical professionals—need to take on leadership roles for many reasons. By demonstrating how technology can improve our lives and our economy, they become visible role models, showing young people why they should devote more attention to STEM, so that these students can make dreams come true—including their own.

You also have an ethical obligation. You may have pursued a technical education because you saw the possibilities of improving people's lives through your work. For example, the designer of a bridge or building, or of a pacemaker, must scrutinize every decision for potential future negative consequences. That's why many engineering programs have students participate in the "Order of the Engineer" ceremony, in which students pledge an oath to use what they have learned for the benefit of humankind. By tradition, each student receives a ring, originally made of steel from the Quebec Bridge, which failed in 1907 and inspired the oath and ceremony.

Obligation of the Engineer

The Obligation of the Engineer (see the appendix) requires that engineers become stewards of nature's vast resources of material and energy. They must use those resources for humanity's benefit; they must practice integrity and fair dealing, tolerance, and respect; and they must participate in nothing but honest enterprises. They must give their skill and knowledge without reservation for the public good and give their utmost to the achievement of these goals.

Although designed for engineers, this statement applies equally to all those with science, technology, engineering, and mathematics education. It applies to you and other technical professionals. In order to give your utmost, you must develop and exercise leadership abilities. This is how you can demonstrate your passion and courage to make a difference.

When we choose professions, most of us think of our heroes. We find people we admire and try to emulate their behavior. They may be captains of industry, powerful political leaders, sports figures, or social leaders. Usually, these heroes are others, not ourselves.

As Parker Palmer wrote in *Let Your Life Speak* (2002), "Our deepest calling is to grow into our own authentic selfhood." But we are not taught to look inward for solutions. The concept that eludes us is that of vocation: the notion, sometimes referred to as a "calling," that within us lies the reason for our existence. It is a matching of the passion that we have with the greater needs of others. Recall Buechner's (1993) assertion mentioned in Chapter 9: "We are called to the place where our deep joy meets the world's deep hunger."

There is joy in unearthing our real selves and identifying how we can contribute to the greater good. We complicate our lives with trivia, but what our parents and grandparents taught us is still true, that the most important things in life are not things. They are honesty, service, compassion, and sharing. Engineers and scientists who develop their leadership abilities can experience the personal satisfaction of sharing valuable skills, the public recognition of making helpful contributions, and the professional rewards of bringing new ideas to life.

William Wulf, former president of the National Academy of Engineering, has said that for U.S. engineers and scientists to compete in a global economy, we must be more innovative. In his words, innovation and creativity are a matter of putting together ideas that already exist but have not been connected before. He also observed that a more diverse scientific and engineering population increases the variety of unique ideas, and therefore the potential for innovation. The power of this idea is great and explains why teams are so important.

The calls for leadership are everywhere; we just need to listen for them. Once we hear and recognize those calls, we must be ready to respond as competent, innovative, authentic leaders who have the technical skills, passion, and courage to make a difference. Because innovation and creativity are leadership skills, and because so many of the problems we face will require technical solutions, technical professionals have the opportunity and obligation to act.

At times, the call is faint; it may come from deep within you, and the meaning may be subtle. We are not used to looking inside for answers, and the activity and noise of daily life can distract us from our calling. You may be struggling with the balance between action and reflection if it still seems foreign. Once you choose to respond, you must be willing to leave comfortable habits and face challenging questions. Life becomes more complicated—but more rewarding and enjoyable—when you answer that call.

Remember, leading is not about position. It is a matter of courage, persuasion, and trust. You can lead from anywhere in your organization, with small groups or large. However, to be effective, you need the passion and confidence that bring the courage to speak up and build excitement among your colleagues. And you must be authentic, true to your own beliefs. One clarifying definition of a leader is someone who will take you to a place you wouldn't go by yourself.

Your roles as a technical leader are to foster innovation by engaging everyone in your organization to bring forth their ideas and to build a safe and welcoming environment for sharing and combining those ideas into creative new approaches. As a professional and leader, you must help others evaluate and prioritize the best ideas and take initiative in acting on them.

Because you are well equipped to evaluate ideas, this kind of responsibility is well within your expertise; you are uniquely qualified to lead innovation. At the same time, you have been trained as a subject matter expert and have become proficient in your technical field. You are likely viewed as the expert in some aspect of your work. That is both an asset and a liability.

To become a leader, you must go beyond your specialized expertise and become a generalist. As Carol Jacobs puts it, "specialize in becoming a generalist." You will have to develop a systems viewpoint, putting your technical expertise into perspective within a broader context. You will need to understand how your decisions affect others and the larger project and be able to empathize with your colleagues, stakeholders, and customers.

The process of developing new ways of thinking about your choices, your decisions, and your environment is much like a fledgling eagle learning to fly. Adult eagles are a joy to watch; they are graceful in flight, impressive in appearance, and skillful at hunting. Fledgling eagles lack all of these characteristics. They are awkward at learning to fly, and they complain loudly when their parents push them to learn, especially about hunting and getting their feet wet. And they really complain when they're pushed out of the nest and told to get their own place. They practice endlessly at learning to fly, hunt, and create their own lives. Eventually, they become experts, growing in maturity and independence, developing self-confidence and becoming adults. Just as eagles do, you will experience awkward moments as you learn to lead. With practice, though, you will also develop into a mature, independent, self-confident leader.

Your success will become an inspiration for others. You are not the only one who wants to succeed, and you benefit even more when others join you in developing their leadership skills. You can probably see who else is ready for flight; engage them in a conversation to take on new roles, and take on some of the challenges that are already in front of you.

We all want to be productive and make a difference. Often your ideas and approaches seem better than those that are being pursued. If you believe you have a better approach, you need to make your ideas visible, to communicate, and to build support among others. With a strong, supportive group, you and your ideas can learn to fly.

Chapter 13 Reflection Questions

1. What 21st-century challenges spark your imagination and passion?
2. What might you do to pursue those interests more fully?

3. How does your vision align with the Grand Challenges cited in this chapter?
4. How do you see your efforts contributing to some of those?
5. What other needs fit your interests?
6. Where does your deep joy meet the world's deep hunger?
7. What callings are lurking deep within you?
8. How might you surface those callings, so that they become part of your roadmap ahead?
9. What is keeping you from taking flight—from going after your dreams?

Broadened Perspectives

The challenges have been laid before you. Are you ready to open up and expand your thinking?

Without a change in direction, we are on the way to a world of increasing population pressure and poverty; growing potential for social and political conflict; official and unofficial warfare; food, water, and energy shortages; worsening industrial, urban, and agricultural pollution; further destruction of the ozone layer and accelerating climate change; continued loss of atmospheric oxygen; and an accelerating reduction of biodiversity. We also run the risk of mega-disasters caused by nuclear accidents and leaking nuclear waste, devastating floods and tornadoes accelerated by climate change, and widespread health problems due to natural catastrophes as well as accumulation of toxins in the soil, air, and water Laszlo 2008). This is not where we want to go.

There is widespread agreement that we are close to a tipping point, and there is a relatively short window of time to right the balance among people, planet, and profit. Malcolm Gladwell (2002) has devoted an entire book to the idea that change happens in the same way that viruses spread, like an epidemic. There are discernable patterns, one of which is the presence of contagious behavior. The second pattern is that small changes lead to proportionately greater results. The third pattern is that changes happen quickly; not much time is needed to build up to a dramatic result. Epidemic change is the result of a dynamic, complex system in which the causes and effects are not easily connected. This kind of dramatic change is what is needed over the next 40 years or sooner to rebalance our own complex world.

Recent discoveries in physics have changed the way we think things work. Our view of motion, time-space, matter, and causality have all been turned upside down by uncertainty, relativity, and chaos. Ray and Anderson (2000) talk of cultural creatives—an emerging culture, about 23% of the U.S. adult population, with a holistic worldview that includes both moral and scientific realms. Their values include learning rather than being entertained; being active in culture and the arts; authenticity in life; and selective and conscious consumerism. They are holistic in their perspectives, and they aspire to have an effect on the culture—whether it is their choice of food or their views on inner growth or their balance between work and play. Are you one of these cultural creatives? How do you describe your worldview? How is that changing as you assess the future and your role in it?

Worldviews affect human behavior, and how we behave affects the world around us. The scientific worldview is comparably significant, and has undergone drastic change during the twentieth century. The physical science ideal of mathematical precision and predictability, as elaborated by Galileo, Newton, and their heirs, underwent an amazing transformation in the twentieth century when Big Bang cosmology substituted an expanding, unstable universe for the Newtonian world machine.

—William McNeill

Our developing leaders were exposed to current leading thinkers and theorists, who write and speak about global challenges and the societal implications of dramatic change. These leaders have been challenged to consider what kind of thinking and leadership is required to bring about solutions for our communities and planet and to ensure a future for the generations to come.

Relevant questions include the following: Am I rethinking old paradigms? Is my perspective being enhanced and expanded, or does it feel as though I am being torn from my foundations? These experiences were meant to be provocative; these emerging leaders were expected to question their mindsets and intentionally reframe some they may have thought previously. This deliberate learning process caused many to broaden their worldviews over time.

Mary Rosen told of her experiences in being provoked to reflect from a totally new place: "I appreciated some of our 'leaderful' discussions that were tied to thinking about our source of inspiration, and the creative essence of who we are. I was greatly impacted by Otto Scharmer's notion of the blind spot and how I might access more of that as I allow myself to let go and let come into thinking about new creations and possibilities. I want my leadership journey to take me in the direction of this great enlightenment. Bill O'Brien's quote 'The success of an intervention depends on the interior condition of the intervener' has significant resonance for me."

Structure of Attention

Effective leadership depends on the quality of attention and intention a leader brings to any situation. Two leaders in the same circumstances can bring about completely different outcomes, depending on the inner space from which each operates. We are often blind to this dimension of leadership and transformational change. This "blind spot" exists in our collective leadership and our everyday interactions. It comes from deep within and is not seen or heard. To be effective, we need to understand the inner space from which we are operating. Otto Scharmer (2009) of MIT identifies four "structures of attention" that result in four different ways of operating (see Figure 14.1). Each field increases in the breadth and depth of perspective.

Scharmer argues that all true innovation in science, business, or society depends on accessing one's inner knowing, not from "downloading." This requires deep listening, observing, and reflection to allow the inner knowing to emerge. He

Field	Micro:	Meso:	Macro:	Mundo:
Structure of Attention	THINKING (individual)	CONVERSING (group)	STRUCTURING (institutions)	ECOSYSTEM COORDINATING (global systems)
Field 1: Operating from the old me-world	Listening 1: Downloading habits of thought	Downloading: Talking nice, politeness, rule-reenacting	Centralized: Machine bureaucracy	Hierarchy: Central plan
Field 2: Operating from the current it-world	Listening 2: Factual, object-focused	Debate: Talking tough, rule-revealing	Decentralized Divisionalized	Market: Competition
Field 3: Operating from the current you-world	Listening 3: Empathic listening	Dialogue: Inquiry, rule-reflecting	Networked: Relational	Dialogue: Mutual adjustment
Field 4: Operating from the highest future possibility that is wanting to emerge	Listening 4: Generative listening	Presencing Collective creativity, flow, rule-generating	Ecosystem Ba (the Japanese word for place)	Collective Presence: Seeing from the emerging whole

FIGURE 14.1 How structure of attention determines the path of social emergence add (Scharmer 2008).

Source: Reprinted with permission of the publisher. From Theory U: Leading from the Future as it Emerges, © 2009 by Otto Scharmer, Berrett-Koehler Publishers, Inc., San Francisco, CA. All rights reserved. www.bkconnection.com

suggests that this is the hard work of leadership; it begins with creating a space that invites others in, then listening carefully to what life is saying. This also requires learning to suspend the voice of judgment that is often tied to the past. When you can let go and create a space of inquiry and wonder, you make an opening for higher future possibilities. Establishing this place of reflection requires rigorous practice in a busy environment and mind, but it holds opportunity for profound creativity to emerge.

Kirk Kanter spoke about his awareness of how reflection time is critical: "I have learned that taking time to reflect on various situations really helps. I am making time to keep a journal that allows me to capture my feelings, my thoughts and possible new ideas, not letting those slip by without understanding them better. I do this daily during my breakfast routines, and that has been helping me open up to new possibilities in my regular workday. The interesting thing about this journal is if I feel like my mind is cluttered, and I can't think straight, writing things down seems to release my emotions or at least allow me to go on with them. It's funny how that works."

Consider the challenges of our century. Are you open to new possibilities, pathways, and practices that can help you take on and solve related issues? Those with vision and creative ideas who also understand the technical foundations of problems will rise as leaders. This is why your role is so important. Your expertise and leadership will be critical to influencing a group through dialogue, generative listening, and other practices that reflect innovative methods.

Throughout history, technical problems were solved by people who had new ideas, based on science and developed through engineering and enabled by technology and mathematics.

Science is about discovery, learning about the fundamental nature of biology, physics, and chemistry at the core of our world. Engineering is about innovation using science, building on fundamental principles, and creating technologies that allow humans to accomplish more than they can without these tools. Technology is what engineers create with their understanding of science. Mathematics is the language used by scientists and engineers who need to communicate complex ideas in a clear, concise way.

Recall the statements of Dr. Joe Ling in Chapter 13: "Environmental *issues* are *emotional*. Environmental *decisions* are *political*. Environmental *solutions* are *technical*." Not only will you need to address the technical issues but you will also need to lead by communicating effectively to those not technically educated to influence the emotional and political aspects. One recent example of this is the work of Dr. John Abraham in examining the issue of global climate change:

In October 2009, Christopher Monckton, a British politician and former newspaper editor, gave a lecture at a Minnesota college in which he claimed there was no scientific evidence of global climate change. After hearing about this lecture, Dr. John Abraham of the School of Engineering at the University of St. Thomas became curious. He found that Monckton had based his case on nine points.

Justification for the claims did not seem sound to Abraham, so he began investigating—using scientific principles and methodology. He contacted many leading world experts for each point, built a sound case, created a presentation, and put it online. The website for this presentation is listed in the bibliography.

Abraham showed how his documentation was based on sound scientific research, yet there may be other facts not yet brought into evidence, and he invited anyone to come forth with any evidence to support Monckton's allegations. He did this without confrontation or defensiveness.

In the process, he gained worldwide support from experts on climate change. Two of those experts have joined Abraham to create a "rapid response team" for anyone who wants to validate claims regarding global climate change. This group has also prepared a comprehensive document for the U.S. Congress, so that legislative leaders will be able to take action based on fact, not myth.

This is an outstanding example of a technical professional standing up, being heard, and taking a leadership role. Abraham and his colleagues are not doing this for fame or fortune, but because it is the right thing to do. This kind of mission requires courage and passion—but, as Abraham knows, it needs to be done.

What other issues need action, and what do you have the passion to do about them?

To make a difference in the world, you must leave the safe surroundings of your technical discipline. You need to recognize the big issues and the implications of your specialty and take on a systems perspective, in which all things are interconnected and interdependent. Ask yourself how your innate ability, technical training, beliefs, and passions put you in a position to solve these problems—or to bring together a team to solve them. What did John Abraham do to become the catalyst for global climate change experts, and how did his initiative lead to widespread awareness of the facts of this issue? Take the time, right now, to view his presentation at the link provided in the appendix.

As a technical professional, you have the technical skills needed to solve problems—and ethical, personal, corporate, and educational obligations to help. Many engineering professions have ethical standards and statements (e.g., National Institute for Engineering Ethics, American Society of Mechanical Engineers; see bibliography). The subject is also important in the sciences (e.g., Physics). Each technical profession has an ethics statement or policy. The best known is the Hippocratic Oath physicians take to "do no harm."

Other science professions have oaths, as do engineering fields. Probably the most succinct is the general engineering oath titled Obligation of the Engineer (see the appendix). This statement calls for engineers to be fair in their dealing, conserve nature's precious resources, and serve the public good. It has become strikingly clear that conserving the earth's limited resources is critical to sustainability. It is the duty of the technical professional to use energy, material, and human resources wisely.

Serving the public good is the core of a society. We can fulfill this obligation only if we see our actions in terms of their effect on everyone, not just ourselves, our community, or our nation.

Ethical Behavior Brings Success

Ethical behavior also brings success. Research shows that corporations with ethical practices do better financially (Weimerskirch 2006). In his presentation, Weimerskirch referred to research cited by James Mitchell in *The Ethical*

Advantage (2001) based on work by John Kotter and James Heskett (1992), who conducted research using a set of characteristics developed to emphasize the concerns of all stakeholders. They identified companies that exhibited those traits, compared them with their industry competitors, and found they had superior stock performance. Their research showed that companies with cultures emphasizing all stakeholders outperformed companies without such cultures. Over an 11-year period, for companies emphasizing concerns of all stakeholders,

- Revenues increased 682% vs. 166%.
- Workforce increased 282% vs. 36%.
- Net income increased 756% vs. 1%.
- Stock price increased 901% vs. 74%.

You've probably heard the term *synergy* used to describe the principle that a whole can be greater than the sum of its parts. Seeing this principle in action is astounding. Peter Senge (2004) discusses it in the context of a learning community "where people continually expand their capacity to create the results they truly desire, where new and expansive patterns of thinking are nurtured, where collective aspiration is set free, and where people are continually learning to see the whole together." When this happens, a group of two or three can produce the output of many because of the collective resonance that inspires the community to new heights.

Collective resonance requires open dialogue among group participants. The same kind of synergy must happen among leaders if we are to create significant change. To move toward generative learning and listening, where a group can operate from its highest future possibility, leaders must effectively transform their conversations from debate to deep inquiry through dialogue. This collective creativity allows for "finding flow," as described in Mihaly Csikszentmihalyi's research on peak experiences. The key, according to Csikszentmihalyi (1997), is to challenge ourselves with tasks that require a high degree of skill and commitment. In short, we can learn the joy of complete engagement through group experience.

Dr. William Wulf (2006) describes creativity as taking two existing ideas and combining them in a new way. As group size increases, so does the number of ideas. Even more powerful, the variety of ideas is enhanced by forming groups from diverse backgrounds—not just ethnic diversity but also age, geographic, social, educational, or any other kind of diversity. The combination of ideas possible from a diverse, creative group is staggering. When put together in an open and stimulating environment, this generates momentum and opportunities for innovation.

If you have worked in a cross-functional team, you may have seen this yourself. Emerging leaders often comment that teams with diverse members generate a greater variety of ideas and creative solutions.

In one of the final dialogue groups in the leadership program, students commented on the rich discussion that resulted from having a diverse group: it had members from Eastern Europe, the Middle East, China, Venezuela, and several regions of North America. These people brought perspectives from different cultures, backgrounds, and experiences, helping expand everyone's worldviews. The dialogue resulted in a "collective resonance" experience that increased the positive energy in the group, moving everyone toward a common purpose.

> One of the leaders said, "This was a truly profound and rich experience to deeply listen to the variety of opinions and perspectives. Everyone was so genuinely interested in learning from each others' ideas, and the time seemed to go so fast. It was great when the group asked for an extra hour to extend the dialogue. I got so much from the group, and it was clear that everyone was feeling that way."

Serving the Public Good

As you reflect on the beliefs that inform your worldview, search for the reasons you chose a technical education and career. For some of our leaders, it was interest in the potential for science to improve the human condition. They had seen new medicines save lives and new materials improve the performance of machines. For some, it was to repair damage done by the misuse of scientific discoveries: synthetic chemicals that negatively affected wildlife, nuclear energy that had been misapplied, and misguided construction projects that devastated neighborhoods.

The initial attraction was often that a subject was interesting, fun, and challenging. As these leaders matured, they saw that they could make a difference and focus science on improving the world. They grew to adopt the notion that they are not the owners of the land, but stewards of the land for future generations.

> Steven Maxwell told his story about how his action learning project inspired him to reach out beyond his immediate workplace. "My action learning project brought balance to my life, meaning to my work, and a servant leadership approach to my style. I feel an amazing sense of accomplishment through volunteering at the Fairview Southdale Emergency Room as well as the American Parkinson's Disease Association. These healthcare-related volunteer activities allow me to have a greater sense of appreciation for the devices that I design and improve at my workplace. Seeing a neuro-stimulation lead or a guidewire, similar to those I have designed, contributing to people leading a happier and healthier life is very powerful. It makes it much easier to go to work and to think innovatively about medical device improvements. Helping others also gives me a sense of satisfaction and appreciation for life that cannot be described in words."

Research by the National Science Foundation shows a strong influence of technical education even among those who eventually pursue careers outside science, engineering, and technology. One report concludes "whether or not they have earned additional degrees or work in an S&E (Science and Engineering) occupation, people who have earned an S&E bachelor's degree report that science and engineering knowledge is important to their job. . . . A majority (52 percent) of those S&E bachelor's degree holders employed as artists, editors, or writers reported that their degree was at least somewhat related to their job" (Lowell and Regets 2006).

Think toward your future and what is possible as you consider what contribution you want to make. What will be your legacy? What are you most excited about creating in the future as you think about the needs that already exist?

Chapter 14 Reflection Questions

1. Consider your worldview; how would you describe it, and what does this tell you about yourself?
2. How does it contribute to your ability to be a leader who will truly make a difference?
3. What does the Obligation of the Engineer mean to you?
4. If you are not an engineer, how does it apply to your discipline?
5. When you ponder the challenges of this century, what emerges for you in terms of feelings, wants, and needs?
6. Think of a time when you had a peak experience of collective resonance. What happened? How did you contribute to the experience?

Collaboration across Borders

We have described many collaborative techniques throughout this book: choosing win–win postures, learning to influence, building strong and healthy relationships, helping others succeed, building learning communities, pursuing dialogue through inquiry, opening ourselves to listen and learn, and so on. We live in such an interconnected world that no single leader can succeed without collaboration or a collective—a team, a neighborhood, a city, an organization, or a country. And we have evidence everywhere of what happens when we choose an adversarial approach to getting things done—conflict, dissatisfaction, loss of motivation, failure, even violence.

As a leader, you cannot be effective alone. Remember that the most creative ideas come from a collaboration of people with diverse backgrounds; include all kinds of groups and people in your networks.

When people are unfamiliar with collaboration, they may be reluctant to share their best ideas. This perception of competitive disadvantage is based on a zero-sum attitude—the belief that, if someone else gains, they lose. In actual experience, the whole gets larger; the whole team gets more. If you're after the same goal, share ideas, even with your competitors. Everyone will win.

There is also a notion that leaders cannot flourish in a competitive culture because their efforts will be undermined by rivals. But collaborative leaders can advocate their viewpoints effectively. What distinguishes them is their ability and commitment to listen to others and to change their minds or build on new contributions. Collaboration can serve as a model for alternative strategies that are based on competition.

Collaborative leadership has three important characteristics. First, collaborators begin any dialogue with nonjudgmental inquiry. They are genuinely curious about events and have no hidden interests. Their aim is to increase individual, group, and organizational effectiveness. Second, they submit their own ideas for critical inquiry by others. They are willing to challenge their own ways of thinking, even discovering the limitations of how they think and act. Third, collaborators believe that shared inquiry can lead to innovation and discovery—even to new worldviews. They are willing to reconsider their own ideas in pursuit of a common good.

As an influence process, collaborative practice asks all stakeholders to participate and fully advocate their views, as well as listen to and respect the views of others. Dialogue that permits open disclosure of each person's beliefs, feelings, and assumptions benefits from the contributions of all community members.

Think about the boundaries or borders that keep you from extending your collaborative practices. You have them in your families and in your social contexts, whether informal or organized groups. Borders within organizations are often based on functions: marketing vs. engineering, sales vs. manufacturing, and so on. Borders across organizational boundaries also divide people: with sister business units or divisions, with competitors, and with alliances, networks, and professional groups, as well as regional territories, such as state and national borders.

Imagine extending the collaborative practices wherever you find yourself, working across borders on behalf of a common good. Is this possible, practical, or probable? We believe it is, and if you are going to succeed as a technical leader in a new world, you must learn how to do this effectively. The world's issues are just too big to leave to competitive forces concerned only about their own agendas, whether political or personal.

> Creating a new future that hinges on widespread accountability and connect-edness requires leaders that convene people in new ways to create conditions where context and practice shifts:
>
> - From a place of fear and fault to one of gifts, generosity, and abundance
> - From a bet on measurement and oversight to one of social fabric and chosen accountability
> - From a focus on advice and predictability of leaders to a focus on evoking the wisdom, capacities and ownership of citizens.
>
> —Peter Block (2008)

Identifying Critical Needs

Most global organizations have cross-functional teams in place to encourage collaboration across boundaries, with inputs affecting product decisions, marketing decisions, and new technologies from functional and regional areas. Many of the emerging leaders we interviewed were engaged in some form of global collaboration, several with virtual teams that rely on technology in order to meet and work together. Virtual teams are becoming more commonplace in large organizations, presenting timing and technical challenges. This requires creative management from collaborative leaders. Some of our leaders have direct reports in various regions across the globe. Orrin Matthews shared his challenges of working with a virtual team:

Orrin's direct reports meet weekly via conference call to conduct their business as a team; hearing each other's issues, problems, and successes and reporting what is most critical and relevant for the full team to be involved with and understand. Each team member shares the latest progress from his or her locale, seeking inputs and support from other team members. Orrin uses these meetings to ensure there is mutual collaboration across the team to suggest ideas, processes, and unique ways of thinking about problems that can end up being synergistic, even though each person's local area or issues are unique. He challenges team members to occasionally lead meetings, creating opportunities for others to share leadership responsibilities. He believes the team needs to be self-directed and able to handle their problems together without his constant guidance. They have begun to take more ownership of their team agendas and the ways in which they want to brainstorm new initiatives that can reduce costs for the whole. Orrin says it is exciting to see his virtual team develop relationships that are strong and healthy, representative of true leaders finding ways to collaborate and win together.

Organizations are becoming adept at building collaborative teams to lead their strategic planning, R&D, and other systemwide efforts. The challenge remains how these teams are facilitated, which methods are used for inquiry and expansive thinking beyond organizational norms. Innovation and new ways of thinking are often sacrificed in the name of time. When people are rushing to complete immediate tasks, they seldom take time to question their methods and means for bringing people together to think and create.

Keith Kutler, in his leadership role as one of four executives, believes in collaboration across the organization. He feels it is a major challenge to involve people in shaping the culture, building loyalty, keeping good people, living in authenticity, and being visible in the organization. He feels everyone needs to give and receive constant feedback and to listen to each other's ideas. He asks for inputs from people in production and includes them in decision making. He is serious about keeping all employees aligned with and informed about organizational goals and business progress. He says it is hard to get collaborative inputs and engagement without having good insight into the major business decisions.

We witness some great collaborative work being done in volunteer organizations, in international relief efforts, in professional societies, and by adjunct faculty. Many of our leaders have found great practice fields in some of these initiatives and have been able to transfer some of their learnings back into their workplaces. Some have found a new passion for their lifelong pursuits.

Ellie Fitzgerald is a native of Sudan. She had an assignment in leadership to identify some critical need, research and validate the need, then propose and implement something she personally would do about it. Ellie studied the problem of poverty in Sudan and proposed that she would travel to Sudan, learn more about some specific aspect of the problem, and establish a nonprofit

(continued on next page)

(continued from page 149)

venture to help. She took two months' leave from her job, did what she proposed, and has established a micro-loan program to help entrepreneurs start businesses. She learned a great deal from this experience, especially how one person can lead change for a cause about which he or she is passionate, and is now even more committed to carry this work forward.

Collaborations

George Ayittey (2005), in his book *Africa Unchained,* points to the importance of small community co-operatives in Africa as the most likely source of success for African nations to emerge from poverty. These organizations need the help of technical professionals. You could get involved globally at a grassroots level and build your leadership skills with organizations such as Engineers Without Borders, Doctors Without Borders, Scientists Without Borders, and other scientific collaborations.

International collaborations build relationships that help technical professionals understand the true nature and scope of the issues abroad, and they help the people in other countries get much-needed technical support and other resources.

Camille George is an engineering professor with a strong interest in developing countries. She joined several other U.S. university faculty members on a project to help people in rural areas of Africa improve their economic and health conditions. Learning from the work of George Ayittey, she and her colleagues helped form an agricultural cooperative to produce and market products based on shea nuts. The leaders of this initiative are Malian women with business acumen who partner with people from the local university and polytechnic to manufacture the equipment they need to automate their business.

Another project in Africa is being carried out by Soul Source Foundation, led by author Elaine Millam. This project partners with women's groups in South Africa and Kenya to keep young girls in school, finishing their secondary education, and to empower women's groups that are formed as a community coalition to set up their own enterprises, earning enough money for each of the women involved to feed her family and help her children go to school. The enterprises include developing a shamba (vegetable garden) and selling produce in local markets, raising chickens, ensuring there is clean water available at schools, and the like. Each enterprise involves small loans given to the women that require their repayment as they progress. The women are assisted in setting up their business plans, being accountable for managing their finances, and ensuring that effective leadership is in place. Empowering women in this way is the most important factor in helping the community overcome cycles of poverty.

Recently, the James A. Baker III Institute for Public Policy at Rice University in Houston, Texas—through its Science and Technology Policy Program, the Transnational China Project, and the Technology, Society and Public Policy Program—hosted an international workshop to identify and examine key issues preventing effective scientific collaboration among researchers working on both sides of the Pacific Ocean. Officials and university scientists from Beijing, Chapel Hill, Hong Kong, Houston, Los Angeles, Nanjing, Shanghai, Singapore, Taipei, Tainan, and Washington, DC, met to discuss ways to facilitate scientific and engineering research. The goal of the workshop was to develop findings and recommendations that describe best practices for collaboration; determine cultural and policy barriers; recommend actions for universities and granting agencies to promote collaboration; and showcase successful collaborations as models for practices in the future. The Baker Institute Policy #42, available through the institute, is a report outlining their successes and findings.

Scientists Without Borders^{SM}, a new global initiative conceived by the New York Academy of Sciences (NYAS) and the United Nations Millennium Project, has unveiled a website (see the bibliography) and database designed to match needs and resources with individuals and organizations working to improve quality of life in the developing world. "Scientists Without Borders is a pioneering initiative that will link and mobilize institutions and people who apply science to improve lives and livelihoods in developing countries," said NYAS President Ellis Rubinstein, who chairs the Scientists Without Borders Advisory Council. "Whether exploring low-cost green energy technologies, improving strategies for sustainable agriculture, or developing tools to prevent and treat disease, science builds knowledge that advances global efforts to enhance health and prosperity."

Collaborations with foreign scientists are the most common in European Union (EU) countries, thanks in part to funding policies. Half of EU research articles had international co-authors in 2007, more than twice the level of two decades ago, according to a major report recently by the US National Science Foundation.

The EU level of international co-authorship is about twice that of the United States, Japan, and India, even though levels in these countries are rising—a sign of the appeal of working across borders. "The phenomenon is across disciplines," says Loet Leydesdorff, a science-metrics expert at the University of Amsterdam. "You can find it everywhere."

Indeed, as Thomas Friedman (2007) has described, in many ways the world *is* flat. Collaboration across borders is happening everywhere. As a technical leader, you should know what these collaborations are about and consider engaging in projects that may provide great learning experiences while broadening your perspectives.

When you look for ways to collaborate with colleagues in other countries, development projects can offer great opportunities. Whatever technical expertise you have to offer, someone wants and needs your help. Don't be overwhelmed by the size and number of problems that need solving in this world. With your leadership and the involvement of others, many problems can be tackled and overcome. It has been done in the past; it can be done again.

The first step after identifying a problem is to assume you can do something about it. Set out to find and remove obstacles that stand between you and your goals. An entrepreneurial colleague who has built several successful businesses told us that, if he had known all the barriers he would face, he might never have started. Focus on your objectives, not on the barriers you may encounter.

Technical professionals often assume they need a lot of information before they begin. They want to be experts in every way possible. But there is never enough information, or enough time to gather all of it. Learn to make decisions based on whatever information you can get in a reasonable time, and then trust your judgment. Know your vision and why it is important to you; then do whatever it takes to realize that vision. It may be risky to act without all the details, but consider the alternative of inaction. In the long run, that may be riskier still.

When you bring up new ideas, naysayers will often say, "It can't be done." Don't let that stop you. Say, "I'll show you," and proceed to do it.

> After successfully forming the School of Engineering, we were told by a very well respected former corporate executive and then leader in the College of Business that "you defied gravity." When asked what he meant, he said that everyone knew it would be impossible to form a School of Engineering here. The forces against it were too great. Who knew? Ignorance can be bliss.
>
> —Author RJB

Another way to develop leadership skills while pursuing your passion is through volunteering locally. You can volunteer and make a difference wherever you are. Not only will you make a difference but you will learn, practice leadership, and become more visible in your community.

> Jerry Johnson told of his many volunteer activities. He is a board member at his daughter's school and actively involved in many of her school projects—events such as plays, nature walks, and special days (such as a Science Day), where he is one of the few fathers to fully participate. He occasionally goes on field trips with the kids and said, "When you volunteer you get to wear many hats." He is a furniture handyman, Scholastic Book sale organizer, chaperone to most events, and "Science Guy." "I truly enjoy being involved with my kids' school activities," Jerry said. He also coaches his daughter's T-ball and hockey teams and is helping her learn

to improve her skating. Beyond his children's activities, like many of his peers he joined Toastmasters. He loves getting feedback from differing points of view and sees it as an excellent way to grow. He is thinking about getting involved in politics, possibly planning to run for mayor in his suburb, and working with the park board committee.

People outside your immediate community may have different goals than you have and may see the world differently. Learn to work with numerous stakeholders affecting or being affected by your leadership style. It may take longer to develop the trust that is required; be open to using your best communication skills to understand the requirements of the community, and seek opportunities for negotiated agreements. Collaborating emphasizes maximum satisfaction for all parties; each exerts both assertive and cooperative behavior. All stakeholders must encourage the mutual expression of their needs and concerns. This is particularly a challenge when working in cultures where there are language barriers.

Regardless of your choices for extending your experience and influence, the world offers opportunity everywhere. As a technical leader, you must find ways to begin practicing collaborative leadership wherever you are, because the world is shrinking and the need for collaboration is growing. Boundaries and borders will continue to blur our vision, cause confusion, and potentially create new problems, but we cannot solve the grand challenges without knowing how to effectively reach across distances and work as partners.

Chapter 15 Reflection Questions

1. How does collaboration become a key practice for you as you exercise your influence and demonstrate your leadership capabilities?
2. What are some of the ways the organizations you are affiliated with are changing today?
3. Where are there deliberate attempts to bring people together across borders and for what purpose?
4. What is emerging in your thought process at this point?
5. Do you see new opportunities for yourself to get involved in collaborative efforts that are about making a difference that is important to you?
6. Are you ready to begin some kind of reflection process that captures your thoughts, ideas, and possibilities as a result of reading this book? How will you do this?

Sustainable Leadership

The importance of sustainability applies not only to our personal journeys but to all aspects of life on earth. In the case of technical professionals, the two are intertwined. You have lots of work ahead—becoming clear about who you are and discovering what you have to contribute. Once your vision is clear and you are on your path, you will need to keep several things in perspective to keep your leadership plan sustainable.

Day-to-day routines and challenges can distract you from your plan. Multiple demands on your attention can cause you to lose focus. These things are normal and to be expected. But leadership development is a lifelong endeavor. You want to bring your best to all dimensions of your life, while influencing others to do the same. Keep the concept of sustainability at the forefront.

Look at Merriam-Webster Dictionary 2008 definition of *sustain:* "to support, hold or bear up from below; bear the weight of, as a structure." Other definitions refer to "endurance, keeping something going, providing for, and nourishing." The meanings are cyclical; the definition of *sustaining* lends itself perfectly to the notion of settling into nature rather than resisting it.

The sustainability movement aims to restore balance to our communities, reconnect what has become separated, and establish sound stewardship of the energy needed to maintain global, social, and economic health. This is only possible with the involvement of passionate leaders who commit to shared principles. As a passionate leader who wants to make a difference, you must understand what part is yours, and why that part is so important.

Concern for the environment is growing worldwide. The movement for sustainability and green manufacturing has been developing for decades and is gaining momentum globally. Sustainability initiatives are underway in Europe (Hong, Kwon, and Roh 2009; Azzone and Noci 1998), Russia (Bartlett and Trifilova 2010), ASEAN (Lopez 2010), South America (Nunes, Valle, and Peixoto 2010), Africa (Ndamba 2010), and North America (Frazier 2008). The strategic green orientation is affecting supply chains globally (Hong et al. 2009) and business/government alliances are developing to promote a low-carbon economy, such as the Green Economy Council in the UK BusinessGreen 2011. The international standards organization ISO has been developing a new international management system standard for energy (ISO 50001). Even before this standard is fully developed and implemented, individual companies have been taking steps to dramatically reduce energy consumption. The 3M Company recently earned its sixth consecutive Energy Star award for its energy conservation efforts in 65 countries (Business Wire 2011).

The sustainability initiative is underway and growing rapidly—and it is calling for leaders. Your technical skills will be needed to reduce energy consumption, minimize carbon footprints, and manage other environmental effects while reducing costs and building alliances to sustain efforts throughout your supply chain. This will test your abilities, passion, and courage. It will require partnering among technical professionals and others in your organization, particularly marketing. Summon your leadership skills and answer the call.

The following are seven principles you can adopt in thinking about sustainability. This list closely reflects the opinions of Jonathon Fink, vice president of Arizona State University's School of Sustainability, and Dieter Helm, an economist from Europe who specializes in energy, water, and infrastructure, as well as others (Hargreaves and Fink 2004).

Sustainable leadership is about self-control. You are only responsible for your own actions. You can only be in charge of what you do and say, as well as how you act in a particular situation. Therefore, learning to be more emotionally and socially intelligent, you can be assured that you know your own thoughts and feelings and how to build healthy and strong relationships with others.

Sustainable leadership creates and preserves sustaining learning. This is learning that matters, that lasts and engages others intellectually, socially, and emotionally. It goes beyond single achievements to create lasting improvements in learning and development that are measureable and noticeable.

Sustainable leadership requires using social and emotional intelligence. You want to take part in something that matters. You want to make a space for yourself and others, so all can become their best selves. You strive to understand others and help others in contributing to a greater cause.

Sustainable leadership sustains the leadership of others. To leave a lasting legacy, develop your leadership and share it with others. Build a community that can carry on if you leave. Recognize that no one leader can do it all; leadership is a distributed necessity and a shared responsibility.

Sustainable leadership cares for the diverse whole, building its capacity. This means awareness of your interdependence on this planet—that whatever you do to the planet or to another human being you do to yourself. If you help build capacity for yourself and others, the whole system benefits.

Sustainable leadership develops rather than depletes human and material resources. It generates incentives that attract the best of leaders. It provides time and resources for leaders to network, to learn from and support each other, and to coach and mentor their followers. Sustainable leadership systems take care of their leaders and help leaders take care of themselves.

Sustainable leadership ensures success over time. Leadership succession is the challenge of letting go, moving on, and planning for your own obsolescence. It reaches beyond individuals to connect the actions of leaders to their predecessors and successors. It preserves purpose and organizational wisdom.

These principles are aptly described in Hank Bolles's set of beliefs: "I believe leaders must act as stewards of the long-term, continued success of the organization. They must have unwavering resolve to do what must be done to produce the best long-term results, committed to the larger intention and purpose of the organization rather than the flippant opinions of Wall Street analysts whose loyalty is limited to the company's ability to meet quarterly earnings projections. This means also acting as trustees for the company's legacy rather than their own. The future success of the company lies in the selection, development, and strategic assignment of key leaders and future leaders. Such leaders, according to Max DuPree, must become 'transformative leaders, guiding their organizations and the people in them to new levels of learning and performance. Leadership thus becomes a process of learning, risk-taking and changing lives.' One true measure of a leader's greatness is the ability to work oneself out of a job, enacting planned obsolescence by grooming successors for more greatness in the next generation."

Other leaders drafted similar principles for applying their leadership in their workplaces. They wanted to ensure that their leadership learning and experiences were sustained.

Developing Others

A common theme that emerged from all of the 50+ interviews conducted with developing leaders was one of "developing others" and serving the broader organization:

Joe Monahan is devoted to helping his direct reports and project team members put together personal visions for themselves, articulate their top goals, and then identify activities that will help them move toward their visions. Joe meets with his people in one-on-one sessions at least once each quarter to check in with them to see how he can support them, to hear their progress reports, and to modify plans as necessary. He wants to help hold them accountable and expects them to take charge of their plans, with his support and guidance as a coach.

Perry Barnes also believes he is responsible as a leader to ensure that his people set goals for their own personal and professional development. He said, "I can only help them with their development if they want me to, and I talk openly how important I think their development is to our company as well as to them. I want them to consider the legacy they want to leave, be dedicated to learning, and stretch themselves to new heights." Perry is full of excitement when his people succeed, and he goes to great lengths to acknowledge their accomplishments and recognize them publicly for their growth.

Patrick Conner also spoke about how important it is to his company and to him to have his people pursue growth and development. He expects his people to identify at least one mentor. "That way I am not the only resource they have for coaching and guiding," he said. Mentors help them stretch themselves to do what they feel is their absolute best. This is not just about attending classes or seminars, he said, but doing something for themselves that can be monitored and documented. He expects his people to seek challenging assignments, identifying areas of need in the organization. He tells them, "I want you to put your hands up to lead some of these project efforts." He also expects them to track their learning and growth, making sure they have effective feedback from those around them.

Ann Jones expects her team members to support each other in their growth and development. She regularly holds team meetings for people to share their achievements and to get feedback and acknowledgment from others on the team. This time of celebration helps the whole team recognize the importance and value of continual growth and learning. She shares her own growth and development with the team as well. At times, the team takes on specific books or articles to read, holding group dialogues or brown-bag sessions that help people make direct applications to their common work activities and share their successes.

Many leaders shared ways of acknowledging people for their efforts in growing, learning, and trying out new ideas. Several shared tactics they use to keep people interested in new pursuits: attending training sessions together; identifying key projects that can be assigned to specific people who are ready for new challenges; and working in partnerships to help people learn new technologies or carry out specific project assignments. Many set up brown-bag sessions where team members identify key topics that they want to learn more about, or learn together about best practices elsewhere, often bringing in associates from other organizations.

What about you? What are your expectations for yourself, and what sustainable practices will keep you on track, continue your learning and growth, and help others do the same?

What you believe influences your reality, so choose to believe that people will grow and take charge wherever possible. When people see that others believe in them, they begin to believe in themselves. Your expectations of and confidence in your team members will help them build confidence in themselves. That will give them energy to achieve their personal and team goals.

Expect technical people to bring in new perspectives, new ideas, and new thinking. Help them exercise their creativity and emotional intelligence. The challenges of global sustainability require the full engagement of every person—beginning with you.

Most of us educated as technical professionals were not trained to take on leadership roles that require the skills of serving others. However, by helping others, we help ourselves and the organizations to which we belong. If you have ever taught a class, been a coach, or mentored someone, you know the rewards. You help others and learn a lot about yourself and your topic.

When asked about future aspirations and plans, many of those we interviewed said they wanted to teach. Others talked about mentoring subordinates or colleagues. The experience is rewarding; it gives you an opportunity to give back with your knowledge and experience and to make a direct impact on someone's life.

> During my career, I have had many mentors. Actually, I didn't realize it at the time. I just knew these were excellent people and they spent time with me, helping me with my career. I didn't know why they would do that, just thought they were fine people. I came to realize that there are a lot of great people and, given the chance, they want to help others. My advice to everyone is to do the same for someone. It could change the person's life.
>
> —Author RJB

We Did It Ourselves

In our roles as mentors, coaches, and teachers, we find that helping others ends up teaching us valuable lessons. It's also valuable practice for sustaining our own leadership principles, beliefs, and possibilities. Our students and direct reports become our teachers. When they identify practices we've never thought of, and they encounter obstacles we haven't experienced, we can still be compassionate listeners and guides.

Although the experiences and challenges are familiar, some things are always unique: the players involved, their attitudes, and their ways of dealing with themselves and others. Keep an open heart and mind as you inquire into their experiences; don't assume that age or experience gives you the best answers. Seeing others discover their own answers can be rewarding. Share your own experiences, but avoid giving advice and direction. Others have their own process and capabilities. Help them discover the truth from within, so they can say "we did it ourselves."

When the best leader's work is done the people say, "We did it ourselves."

—Lao Tsu

Once again, reflection is a key process for learning and sustaining ourselves. How often do you stop and think, why am I doing what I'm doing? Is it for money or recognition, or because you just can't see a way to pursue what you really want? Are you finding pleasure and personal satisfaction in your work? Do you feel your work is contributing to some larger goal that is important to you? Or do you hate what you're doing and wish it were something else, but just don't know how to get out? Are you asking yourself with regard to your work and avocation, for the sake of what?

A colleague recently asked, "Where am I going, and why am I in a handbasket?" A couple of coaching sessions helped him find answers to some of his frustrations. He had not given himself time and energy to think through his wants and needs and had allowed himself to get stuck. He had no experience of what it means to sustain himself as a leader. Once he became aware of his situation, he had the means to begin again. All he needed was for someone to help in the discovery process, and once again he was out of the handbasket—making plans and taking charge of his destiny.

Another colleague believed early in his career that he wanted to work only in industries that he perceived had potential for good. His first professional employment was in the appliance industry, making toasters, coffeemakers, and steam irons. Next was in the electronics industry, making components for TV and radio communications to provide news and entertainment. This eventually was not enough, and he entered the medical device business to save and improve lives. Then he perceived that the knowledge business helping companies adopt new ideas could add more value. And, most recently, he entered the education business to help working adults grow and achieve their potential. At each step in his career, he responded to reflection and the desire to contribute something for the betterment of society. It has been a lifelong journey, and the future continues to open new and exciting possibilities.

Young adults are looking for purpose in their lives. In their world, possibilities do not seem endless. The prospect of having a lower standard of living than their parents is not just possible but likely, and what were once seen as boundless natural resources are clearly limited. A growing world population and increasing standards of living in developing countries offer the promise of even greater consumption and pressure on the world's resources.

By contrast, young adults at the beginning of the 20th century did not have much in terms of wealth or luxuries. Despite the abundant natural resources, that generation practiced frugality, delayed gratification, and modest living. They were not large consumers of resources. By today's standards, they would be considered poor. Yet they didn't perceive themselves to be poor. They had jobs, a family life, a church community, vacations, and fun. They lived productive and full lives.

Creative and innovative ideas led to the products and conveniences developed during the 20th century. Mass production and inexpensive natural resources produced a multitude of products that could be sold at low cost, affordable to a rapidly growing middle class. Making these products created jobs, which provided income, which supported more consumption. The cycle of production, consumption, and growth continued. This process not only created benefits to society but also laid the foundation for the technical problems we face today.

Solutions to today's problems will also come from innovative ideas. The challenges are even greater with sustainability the critical goal. Many solutions will be based on technology, so the skills of technical professionals are essential. A greater challenge will be changing habits, because they reflect beliefs. Technical professionals will also need to understand the emotional and political aspects of the issues and take responsibility for debunking myths and revealing truths. Your mission is to have the data first, then tell the story in a way everyone can understand.

Many developing leaders shared specific ways they nourish themselves and sustain their leadership capacity. Some have become champions for causes outside their work: in their children's schools, their communities, and their churches. They needed to be active, to be visible, to share their ideas, to connect with others, to influence the broader community, and to network with others to expand their perspectives on leadership. Some ran for political office, and some were active in industry groups; one created a new professional society. Many chose to help children—through scouting, coaching sports, giving private lessons. Some volunteered at their local food shelf or hospital. One chose to become a mentor to a prisoner.

In all these situations, the leaders shared the deep gratitude and learning that emerged from these experiences. They recognized clearly that people want to belong, to help each other, to learn from each other. They noted that their leadership modeling for others becomes contagious. It sparks new possibilities for others to experience the "deep joy that meets the world's deep hunger." Most of these experiences taught the leaders that there are many ways to serve their communities and their world. These experiences also create collaborative networks that inspire, teach, and build confidence. This is sustainable leadership.

Caring for ourselves is often at the bottom of our priorities, especially if we are high achievers. Too many truly effective leaders have fallen into this trap and have found themselves at some point "totally burned out." This is especially a phenomenon in high-performing organizations, where there is never enough time or resources to get everything done. Be conscious of the need for revitalizing yourself.

Your Leadership Development Plan

Within your leadership development plan, make sure you have specific activities for renewing and restoring yourself at regular intervals. Even with high demands, plan for a vacation or at least some time off—whatever you truly enjoy. This is not selfishness, but a necessary practice to maintain your energy and vitality. Unless you learn to help yourself, you can hardly help others. Make a list of sustaining activities and put them on your calendar—in permanent ink.

Chapter 16 Reflection Questions

1. What practices or principles will guide you in becoming a sustainable leader?
2. How will each specifically help you?
3. Consider taking on a mentoring role or being mentored yourself. What do you feel are the most important things to consider in a mentoring relationship?
4. As you consider "self-care," how will you ensure you are giving yourself opportunities for renewing and revitalizing yourself? What is your plan?
5. What ways do you want to make sure you are reaching out in your community to sustain your learning and leading?

The world needs you. There are big problems to solve in this century. As a technical professional, you have an important responsibility to help find better answers. You have the education, training, experience, and critical thinking skills needed to innovate and lead the creation of sustainable solutions. You cannot do this alone, so you also need to develop the leadership skills, attitudes, and actions necessary to engage others in the process.

Many of the critical issues we face will require technical solutions. However, as Dr. Joe Ling has pointed out, there are also emotional and political aspects to consider. To be effective in engaging others to assist, you need to develop an understanding and a mastery of these aspects as well. Your continuing learning action plan should include knowledge of others' motivations and behaviors, building relationships and becoming a servant leader.

Before you can lead others, you need to know yourself thoroughly. Understanding your true beliefs and passions will require reflection and exploration. This will be an adventure, and you won't know what you'll find until you've ventured to places in your mind that may not have been visited recently. You may discover things you don't like; you will likely also find some beliefs and passions that will bring you great joy. With this newfound self, you are prepared to develop your lifelong learning plan and roadmap and to set out on your leadership journey.

Where this journey will take you is a mystery at the beginning. It will probably lead to paths you didn't know existed:

- You may discover great needs in your work organization that match your passion and step up to leadership with fresh ideas, building the relationships you need as you go and tapping into existing relationships that provide support.
- You may find needs in your community that fit other passions, such as helping youngsters develop a love of learning, especially of science, technology, mathematics, and engineering. This need may be in terms of technical depth or of building technological literacy in those with other passions.
- Perhaps your role is to debunk misconceptions about science and technology. Notions that are unfounded but often repeated become assumed truths; sometimes they are the result of deliberate misinformation. Technical experts can respond with persuasive facts.
- As members of professional associations, you have the opportunity to take leadership roles and support the goals of those organizations. You may even help organizations reevaluate and realign their goals.

■ With your technical skills, understanding of your passions, and newfound leadership skills, you may delve into public policy. Few members of our legislative bodies have technical backgrounds. To write good public policy, they need your expertise as an advisor—or perhaps as an elected peer.

Everything is changing rapidly. You may discover that you need to continue your education. There are many ways, formal and informal, to learn. Take advantage of those that fit you best.

Now that you know how to reflect and see the benefits, you will need to make time to continue reflection. Even your beliefs and passions will change, and you need to stay tuned to yourself.

Above all, think globally. The technical issues facing the world are global issues. Some may be even more critical in developing areas than where you live. Are you aware of these issues? Can you understand what they mean to the people living in those areas? What can you do to help? Are you doing it? Perhaps there are opportunities through your community, your religious organization, your professional association, or a local college or university to become engaged.

As a technical professional, you have the fundamental skill of critical thinking. Use your skill, specific technical knowledge, and ability to lead to have a positive impact in the world.

Think back to when you were young. Why did you pursue the path you have taken? Was there some large, underlying goal of helping people, solving technical problems, or other form of making a difference? Now that you have all the tools identified, revisit your personal goals and answer the question *"What are you going to do to really make a difference?"*

I. Obligation of the Engineer

I am an Engineer. In my profession I take deep pride. To it, I owe solemn obligations.

Since the Stone Age, human progress has been spurred by the engineering genius. Engineers have made usable nature's vast resources of material and energy for Humanity's [Mankind's] benefit. Engineers have vitalized and turned to practical use the principles of science and the means of technology. Were it not for this heritage of accumulated experience, my efforts would be feeble.

As an Engineer, I pledge to practice integrity and fair dealing, tolerance and respect, and to uphold devotion to the standards and the dignity of my profession, conscious always that my skill carries with it the obligation to serve humanity by making the best use of Earth's precious wealth.

As an Engineer, [in humility and with the need for Divine guidance,] I shall participate in none but honest enterprises. When needed, my skill and knowledge shall be given without reservation for the public good. In the performance of duty and in fidelity to my profession, I shall give the utmost.

Note: Brackets [] indicate the original wording of the Obligation. Either wording is acceptable, but new certificates have the newer wording.

www.order-of-the-engineer.org/?page_id=6

II. Assessment Tools

a. Zenger-Folkman 360 Feedback Assessment: www.zfco.com/pdf/ePressKit.pdf

b. Strengths Finder Assessment: www.strengthsfinder.com/

c. Signature Strengths: www.authentichappiness.sas.upenn.edu/default.aspx

d. EQ In Action Profile: www.learninginaction.com

e. The Leadership Circle 360 Assessment: www.theleadershipcircle.com/

f. Social Styles

    i. Wilson Learning: http://wilsonlearning.com/capabilities/individual_effectiveness/social_styles/

    ii. Social Style Assessment: http://coursesite.uhcl.edu/BPA/robinson/images/socialstyles.pdf

III. Engineering Accreditation Commission Criterion 3

From Criteria for Accrediting Engineering Programs, 2011–2012 Accreditation Cycle, Engineering Accreditation Commission of ABET, October 30, 2010.

The program must have documented student outcomes that prepare graduates to attain the program educational objectives. Student outcomes are outcomes (a) through (k) plus any additional outcomes that may be articulated by the program.

a. an ability to apply knowledge of mathematics, science, and engineering

b. an ability to design and conduct experiments, as well as to analyze and interpret data

c. an ability to design a system, component, or process to meet desired needs within realistic constraints such as economic, environmental, social, political, ethical, health and safety, manufacturability, and sustainability

d. an ability to function on multidisciplinary teams

e. an ability to identify, formulate, and solve engineering problems

f. an understanding of professional and ethical responsibility

g. an ability to communicate effectively

h. the broad education necessary to understand the impact of engineering solutions in a global, economic, environmental, and societal context

i. a recognition of the need for, and an ability to engage in life-long learning

j. a knowledge of contemporary issues

k. an ability to use the techniques, skills, and modern engineering tools necessary for engineering practice.

IV. Development Plan Sample

Creating Your Development Plan

Reflect-----Assess-----Understand-----Interpret-----
Development planning will ensure that your current insights will actually help you overcome your leadership challenges and achieve your goals. This planning step will help you use what you have learned in your journey to address your needs, wants, and desires and leverage your strengths.

1. Write your goals in the first column.

■ Target no more than three goals for addressing development needs and goals for leveraging your strengths. Creating more goals than that will diminish your ability to focus on any of them.

■ Be as specific as possible in writing your goals. For example, "improve my ability to negotiate roles and responsibilities in the merged IT organizations" is much more useful than "improve my negotiation skills."

2. In the second column, explain why each goal is important to you.

■ You should only be setting goals that will build your capability to address critical leadership challenges and/or to advance your career as a leader. Having a clear sense of why your goals are important should help you maintain your focus on achieving them.

3. In the third column, explain the actions you will take to achieve each goal.

■ *What you will do or change now.* These are the specific things you can do or start immediately to begin achieving your goals. They may involve, for example, further self-awareness exercises or resolutions to immediately change specific leadership behaviors.

■ *Other development activities.* These are the longer-term activities and changes that you plan to make. They may involve books to read, courses to take, projects to launch, a mentoring relationship to establish, and the like.

■ Seek to use exercises, daily practices, special projects and activities, and self-study resources for developing each of the leadership capabilities and skills you have identified.

4. In the fourth column, explain what support you need from others, in terms of time and/or resources: specifically, what will you ask them to provide?

■ *Note:* Having a coach you trust and respect may be the single most important factor in achieving your development goals. This person may support you with regular check-in meetings or periodic feedback on an as-needed basis. He or she may also be able to provide resources such as time off for training or access to experts. The coach may be your manager or someone else with whom you have a good relationship and who can help you reach your goals.

5. Finally, through the obstacles you will encounter in reaching your goals and what you can do to avoid or overcome them. Two of the most common are the following:

■ *Lack of time.* If this is a problem, you might want to identify certain time-efficient practices you can use on an everyday basis. You might also try setting aside a regular block of daily, weekly, or monthly time that you will spend working toward your goals.

■ *Lack of support.* The best remedies in this case are to identify a coach, as previously discussed, and/or colleagues who have similar goals and want to partner in achieving them.

MY DEVELOPMENT PLAN

Development Goals	Why the Goals Are Important	Actions I Will Take to Achieve Them	Support Needed

Obstacles to Reaching Goals	Steps to Overcome Them

V. Values Map (Hall 1995)

VI. Other Related Sites

Dr. John Abraham's presentation: www.stthomas.edu/engineering/jpabraham/

Fortune 500 companies of 1995: http://money.cnn.com/magazines/fortune/fortune500_archive/full/1955/index.html

ISO 50001 Energy Management Standard: www.iso.org/iso/hot_topics/hot_topics_energy/energy_management_system_standard.htm

VII. Contact Information

Ronald J. Bennett, PhD: rjbennett@stthomas.edu

Elaine Millam, EdD: emillam@work-wise.org

VALUES MAP

PHASES ▶	Phase I SURVIVING		Phase II BELONGING		Phase III SELF-INITIATING		Phase IV INTERDEPENDING	
WORLD-VIEW ▶	The world is a mystery over which I have no control.		The world is a problem with which I must cope.		The world is a project in which I want to participate.		The world is a mystery for which we care on a global scale.	
STAGES ▶	1: SAFETY	2: SECURITY	3: FAMILY	4: INSTITUTION	5: VOCATION	6: NEW ORDER	7: WISDOM	8: WORLD ORDER
GOALS ▶	Self Interest/ Control Self Preservation Wonder/Awe/ Fate	Physical Delight Security	Family/ Belonging Fantasy/Play Self Worth	Belief/Philosophy Competence/ Confidence Play/Recreation Work/Labor	Equality/Liberation Integration/Wholeness Self Actualization Service/Vocation	Art/Beauty Being Self Construction/New Order Contemplation Faith/Risk/Vision Human Dignity Knowledge/Insight Presence	Intimacy/ Solitude Truth/Wisdom	Ecority Global Harmony
MEANS ▶	Food/Warmth/ Shelter Function/ Physical Safety/Survival	Affection/ Physical Economics/Profit Property/Control Sensory/Pleasure/ Sexuality Territory/ Security Wonder/ Curiosity	Being Liked Care/Nuture Control/Order/ Discipline Courtesy/ Hospitality Dexterity/ Coordination Endurance/ Patience Equilibrium Friendship/ Belonging Obedience/Duty Prestige/Image Rights/Respect Social Affirmation Support/Peer Tradition	Achievement/ Success Administration/ Control Communication/Info Competition Design/Pattern/Order Duty/Obligation Economics/Success Education/ Certification Efficiency/Planning Hierarchy/Order Honor Law/Rule Loyalty/Fidelity Management Membership/ Institution Ownership Patriotism/Esteem Productivity Reason Responsibility Rule/Accountability Technology/Science Unity/Uniformity Workmanship/Art/ Craft	Adaptability/Flexibility Authority/Honesty Congruence Decision/Initiation Empathy Equity/Rights Expressiveness/Joy Generosity/Compassion Health/Healing Independence Law/Guide Limitation/Acceptance Mutual Obedience Quality/Evaluation Relaxation Search/Meaning/Hope Self Assertion Sharing/Listening/Trust	Accountability/Ethics Collaboration Community/Supportive Complementarity Corporation/New Order Creativity Detachment/Solitude Discernment Education/Knowledge Growth Expansion Intimacy Justice/Social Order Leisure Limitation/Celebration Mission/Objectives Mutual Accountability Pioneerism/Innovation Research Ritual/Communication Simplicity/Play Unity/Diversity	Community/ Personalist Interdependence Minessence Prophet/Vision Synergy Transcendence/ Solitude	Convivial Technology Global Justice Human Rights Macro-economics
LEADER-SHIP STYLE ▶	AUTHORITARIAN Oppressive dictator with followers who are totally dependent.	PATERNALIST Benevolent paternalist with followers who are dependent and obedient.	MANAGER Efficient manager with followers who are loyally devoted to the organization.	FACILITATOR Listener, clarifier, and supporter with followers who are also listeners, clarifiers, and supporters.	COLLABORATOR Facilitator, producer, and creator with active peer participation.	SERVANT Interdependent administrator with collegial participation.	VISIONARY Liberator with a global network of peer visionaries.	

BIBLIOGRAPHY

3M Earns ENERGY STAR Sustained Excellence Award For Industry-Leading Seventh Consecutive Year. *Business Wire.* April 12, 2011.

Aritzeta, Aitor, Stephen Swails and Barbara Senior. January, 2007. Belbin's Team Role Model. Development, Validity, and Applications for Team Building. *Journal of Management Studies,* Volume 44, Issue 1, pp. 96–118.

Ashforth, Blake. 2001. *Role Transitions in Organizational Life: An Identity-based Perspective.* New York: Lawrence Erlbaum Books.

Ayittey, George. 2005. *Africa Unchained: The Blueprint for Africa's Future.* New York: Palgrave Macmillan.

Azzone, Giovanni and Giuliano Noci. 1998. Identifying Effective PMSs for the Deploying of "green" Manufacturing Strategies. *International Journal of Operations & Production Management* 18(4): 308–35.

Barsh, Joanna, Susie Cranston and Rebecca Craske. 2008. Centered leadership: How talented women thrive. *McKinsey Quarterly* 4: 35–39.

Bartlett, Dea and Anna Trifilova. 2010. Green technology and eco-innovation: Seven case-studies from a Russian manufacturing context. *Journal of Manufacturing Technology Management* 21(8): 910–29.

Bennett, Ronald J., and Elaine R. Millam. 2011a. Educating manufacturing leaders: Creating an industrial culture for a sustainable future. *ASEE Annual Conference and Exposition,* Vancouver, BC, June 26–29.

Bennett, Ronald J., and Elaine R. Millam. 2011b. Developing leadership attitudes and skills in working adult technical graduate students: Research interview results with alumni. *ASEE Annual Conference and Exposition,* Vancouver, BC, June 26–29.

Bennett, Ronald J., and Elaine R. Millam. 2011c. Transforming cultures in industry: Building leadership capabilities for working adult graduate students. *ASEE Annual Conference and Exposition,* Vancouver, BC, June 26–29.

Bennett, Ronald J., and Elaine R. Millam. 2012. Leadership education for engineers: Engineering schools' interest and practice. *ASEE Annual Conference and Exposition,* San Antonio, TX, June 2012.

Bennis, Warren. 1989. *On Becoming a Leader.* New York: Addison-Wesley.

Bennis, Warren. 2009. *On Becoming a Leader: The Leadership Classic.* New York: The Perseus Book Group.

Bennis, Warren, and R. J. Thomas. 2002. *Geeks and Geezers: How Era, Values, and Defining Moments Shape Leaders.* Boston: Harvard Business School Press.

Bennis, Warren and Burt Nanus. 2003. *Leaders: Strategies for Taking Charge.* New York: Harper Paperbacks.

Block, Peter. 1991. *The Empowered Manager: Positive Political Skills at Work.* San Francisco: Jossey Bass.

Block, Peter. 2008. *Community: The Structure of Belonging.* San Francisco: Berrett Koehler.

Bolman, Lee G., and Terrence E. Deal. 1995. *Leading with Soul.* San Francisco: Jossey Bass.

Bolton, Robert and Dorothy Grover Bolton. 1984. *Social Style/Management Style: Developing Productive Work Relationships.* New York: Amacom.

Boyatzis, Richard, Daniel Goleman and Annie McKee. 2004. *Primal Leadership: Realizing the Power of Emotional Intelligence.* Boston: Harvard Publishing Press.

Boyatzis, Richard, Annie McKee and Fran Johnston. 2008. *Becoming the Resonant Leader: Develop Your Emotional Intelligence, Renew Your Relationships, Sustain Your Effectiveness.* Boston: Harvard Business School Press.

Buckingham, Marcus and Donald Clifton. 2001. *Now Discover Your Strengths.* New York: Free Press.

Buechner, Frederick. 1993. *Wishful Thinking: A Seeker's ABC.* San Francisco: Harper Row.

Buffet, Warren. 2005. *Wall Street Journal,* November 12, p. A5A.

BusinessGreen staff. Government Launches Green Economy Council. *BusinessGreen*, Feb 16, 2011.

Collins, Jim. 2001. Level 5 Leadership: The Triumph of Humility and Fierce Resolve. *Harvard Business Review* (January). http://hbr.org/2001/01/level-5-leadership-the-triumph-of-humility-and-fierce-resolve/ar/pr.

Constable, George and Bob Somerville. 2005. *A Century of Innovation.* Washington, DC: Joseph Henry Press.

Cooper, Robert K., and Ayman Sawaf. 1997. *Executive EQ: Emotional Intelligence in Leadership and Organizations.* New York: Berkley.

Czikszentmihalyi, Mihaly. 1997. *Finding Flow: The Psychology of Engagement with Everyday Life.* New York: Basic Books.

Day, Christine R. 2000. *Discovering Connections.* Dearborn: University of Michigan Press.

Druskat, Vanessa Urch and Steven B. Wolff. 2001. Building the emotional intelligence of groups. *Harvard Business Review* (March). http://hbr.org/2001/03/building-the-emotional-intelligence-of-groups/ar/1.

Elkington, John. 1999. *Cannibals with Forks.* Oxford, UK. Capstone Press.

Ethics in Science. www.files.chem.vt.edu/chem-ed/ethics/.

Frazier, Maya. 2008. Who's in charge of green? *Advertising Age,* June 9.

Friedman, Thomas. 2007. *The World Is Flat 3.0: A Brief History of the Twenty-First Century.* New York: Picador Reading Press.

George, Bill. 2003. *Authentic Leadership: Rediscovering the Secrets to Creating Lasting Value.* San Francisco: Jossey Bass.

George, Bill. 2007. *True North: Discover Your Authentic Leadership.* San Francisco: Jossey Bass.

Gladwell, Malcolm. 2002. *The Tipping Point: How Little Things Can Make a Big Difference.* Boston: Little, Brown.

Goleman, Daniel, Richard Boyatzis and Annie McKee. 2002. *Primal Leadership: Realizing the Power of Emotional Intelligence.* Boston: Harvard Business School Press.

Greenleaf, Robert K. 1991. *Servant Leadership: A Journey into the Nature of Legitimate Power and Greatness.* Indianapolis: Paulist Press.

Hagberg, Janet. 2002. *Real Power: Stages of Personal Power in Organizations.* 3rd ed. Minneapolis: Sheffield.

Hall, Brian P. 1995. *Values Shift: A Guide to Personal and Organizational Transformation.* New York: Twin Lights.

Hargreaves, Andy and Dean Fink. 2004. The seven principles of sustainable leadership. *Educational Leadership* 61(7): 8–13.

Hesselbein, Frances, Marshall Goldsmith and Richard Beckhard, eds. 1996. *The Leader of the Future.* San Francisco: Jossey Bass.

Hesselbein, Frances and Marshall Goldsmith. 2006. *The Leader of the Future 2: Visions, Strategies, & Practices for the New Era.* San Francisco: Jossey Bass.

HBR OnPoint Collection. 2000. *Motivating Others to Follow.* Boston: Harvard Business Review Press.

Hirsch, Sandra and J. Kise. 1994. *Work It Out: Clues to Solving People Problems at Work.* New York: Davies-Black.

Hong, Paul, He-Boong Kwon and James Jungbae Roh. 2009. Implementation of strategic green orientation in supply chain. *European Journal of Innovation Management* 12(4): 512–32.

Hudson, Frederic M. 1999. *The Adult Years: Mastering the Art of Self-Renewal.* San Francisco: Jossey Bass.

Jordan, Peter and Neal Troth. 2004. Emotional Intelligence in Organizational Behavior and Industrial-Organizational Psychology, Chapter in *Science of Emotional Intelligence: Knowns and Unknowns* by Gerald Matthews, Moshe Zeidner and Richard Roberts. New York: Oxford University Press.

Kegan, Robert and Lisa Lahey. 2001. *How the Way We Talk Can Change the Way We Work.* San Francisco: Jossey Bass.

Kolb, D. A. 1981. *Learning Style Inventory: Self-Scoring Inventory and Interpretation Booklet.* Boston: McBer & Company.

Kolb, David A. 2007. *Learning Style Inventory.* New York: Hay Group Learning.

Kotter, John P. 1990. *A Force for Change: How Leadership Differs from Management.* New York: Free Press.

Kotter, John P., and Dan S. Cohen. 2002. *The Heart of Change: Real-Life Stories of How People Change Their Organizations.* Boston: Harvard Business School Press.

Kotter, John P., and James L. Heskett. 1992. *Corporate Culture and Performance.* New York: Maxwell Macmillan International.

Kotter, John P., and H. Rothgeber. 2006. *Our Iceberg Is Melting: Changing and Succeeding under Any Conditions.* New York: St. Martin's Press.

Kouzes, James and Barry Z. Posner. 1987. *The Leadership Challenge: How to Get Extraordinary Things Done in Organizations.* San Francisco: Jossey Bass.

Kouzes, James and Barry Z. Posner. 1993. *Credibility: How Leaders Gain and Lose It, Why People Demand It.* San Francisco: Jossey Bass.

Laszlo, Ervin. 2008. *Quantum Shift in the Global Brain: How the New Scientific Reality Can Change Us and the World.* Rochester, VT: Inner Traditions.

Leonard, George. 1992. Mastery: The Keys to Success and Long-Term Fulfillment. New York: Penguin.

Levine, Barbara Hoberman. 2000. *Your Body Believes Every Word You Say.* Fairfield, CT: WordsWork.

Loehr, Jim and Tony Schwartz. 2004. *The Power of Full Engagement.* New York: Free Press.

Leonard, George. 1992. Mastery: The Keys to Success and Long-Term Fulfillment. New York: Penguin.

Lopez, Jennifer. 2010. ACCA MaSRA 2010: Sustainability reporting trend picks up pace. *The Edge Malaysia,* November.

Lowell, B. Lindsay and Mark Regets. 2006. *A Half Century Snapshot of the STEM Workforce, 1050-2000.* Commission on Professionals in Science and Technology, STEM Workforce Data Project, White Paper No. 1. Washington, D.C: Commission on Professionals in Science and Technology.

Mayer, Peter, Marc Brackett and Jack Salovey. 1997. *Emotional Intelligence: Key Readings on the Mayer and Salovey Model.* New York: Dude.

Millam, Elaine R., and Ronald J. Bennett. 2004. Beyond professionalism to leadership: Leveraging leadership for a lifetime. *ASEE Annual Conference Proceedings,* Salt Lake City, UT, June 20–23.

Millam, Elaine R., and Ronald J. Bennett. 2011. Developing leadership capacity in working adult women technical graduate students: Research interview results with alumni. *ASEE Annual Conference Proceedings,* Vancouver, BC, June 26–29.

MIT Technology Review. S. Rosser and M. Taylor. January/February 2008. Why Women Leave Science. Retrieved from http://technologyreview.com/article/21859.

Mitchell, James. 2001. *The Ethical Advantage.* Minneapolis MN: University of St. Thomas Center for Ethical Business Cultures.

Murray, William. 1993. *Relationship Selling.* Minneapolis MN: Eagle Learning Center.

National Academy of Engineering. 2005. *Rising above the Gathering Storm: Energizing and Employing America for a Brighter Economic Future.* Washington, DC: National Academies Press.

National Academy of Engineering. 2008. *Changing the Conversation.* Washington, DC: National Academies Press.

National Academy of Engineering. National Academies NEWS. Feb 15, 2008. http://www8.nationalacademies.org/onpinews/newsitem.aspx?RecordID=02152008

National Institute for Engineering Ethics, www.niee.org/codes.htm, has links to many engineering society ethics statements. It also has a collaboration with the Center for Study of Ethics in the Professions, which lists 850 ethics codes for a variety of professional associations.

National Science Foundation. 2003. Division of Science Resources Statistics, Scientists and Engineers Statistical Data System (SESTAT). Figure 5. S&E bachelor's degree holders in management jobs, by years since degree: 2003.

Ndamba, Rodney. 2010. Should corporate sustainability be philanthropic? *Financial Gazette* (Harare).

Nunes, K. R. A., R. Valle and J. A. A. Peixoto. 2010. Automotive industry sustainability reports: A comparison of Brazilian and German factories. *VI Congresso Nacional de Excelencia em Gestao* (ISSN 1984-9354).

Palmer, Parker J. 2002. *Let Your Life Speak: Listening for the Voice of Vocation.* San Francisco: Jossey Bass.

Piaget, Jean. Translated by Marjorie and Ruth Gabain. 2008. *The Language and Thought of the Child.* New York: Routledge.

Pink, Daniel H. 2005. *A Whole New Mind.* New York: Berkley.

Quinn, Robert E. 1996. *Deep Change: Discovering the Leader Within.* San Francisco: Jossey Bass.

Ray, Paul and Sherrie Anderson. 2000. *The Cultural Creatives: How 50 Million People Are Changing the World.* New York: Harmony Books.

Rosen, Robert and Lisa Berger. 2002. *The Healthy Company: Eight Strategies to Develop People, Productivity and Profits.* New York: Tarcher.

Scharmer, Otto. 2009. *Theory U: Leading from the Future as It Emerges.* San Francisco: Berrett Koehler.

Seligman, Martin P. 2003. *Authentic Happiness: Using the New Positive Psychology to Realize Your Potential for Lasting Fulfillment.* New York: Free Press.

Senge, Peter, C. Otto Scharmer, Joseph Jaworski and Betty Sue Flowers. 2004. *Presence: Human Purpose and the Field of the Future.* Cambridge, MA: Society for Organizational Learning.

Sullivan, William M., and Matthew S. Rosin. 2008. *A New Agenda for Higher Education.* San Francisco: Jossey Bass.

Tufte, Edward R. 1983. *The Visual Display of Quantitative Information.* Cheshire, CT: Graphics Press.

Vaill, Peter. 1996. *Learning as a Way of Being: Strategies for Survival in a World of Permanent White Water.* San Francisco: Jossey Bass.

Weimerskirch, Arnold. 2006. Presentation on social responsibility. University of St. Thomas, St. Paul, MN.

Weiss, H. M. and R. Cropansano. 1996. Affective events theory: A theoretical discussion of structure, causes and consequences of affective experiences at work. *Research in Organizational Behavior*, 18. pp.1–74.

Wilson, Larry. 1987. *Changing the Game: The New Way to Sell.* New York: Simon & Schuster.

Wilson, Larry. 1994. *Stop Selling: Start Partnering.* Essex Junction, VT: Oliver Wight.

Win the Energy Challenge with ISO 5001. *ISO 50001 Management System Standard for Energy.* www.iso.org/iso/iso_50001_energy .pdf.

Wulf, William. January 2006. Engineering Minnesota's Future keynote speech, http://stream. stthomas.edu/view.htm?id=engineeringWilliam WulfLecture.

Zander, Rosamund Stone and Benjamin Zander. 2000. *The Art of Possibility.* New York: Penguin Books.

Zenger, J. H., and J. Folkman. 2002. *The Extraordinary Leader: Turning Good Managers into Great Leaders.* New York: McGraw-Hill.

Other Useful References

Axtell, Roger E. 1998. *Gestures: The Do's and Taboos of Body Language around the World.* New York: John Wiley.

Bell, Chip R. 1998. *Managers as Mentors: Building Partnerships for Learning.* San Francisco: Berrett Koehler.

Burns, James McGregor. 1978. *Leadership.* New York: Harper & Row.

Charan, Ram. 2008. *Leaders at All Levels.* San Francisco: Jossey Bass.

Charan, Ram, Stephen Drotter and James Noel. 2001. *The Leadership Pipeline.* San Francisco: Jossey Bass.

Christensen, Clayton M., and Michael E. Raynor. 2003. *The Innovator's Solution.* Boston: Harvard Business School Press.

Christensen, Clayton M., Scott D. Anthony and Erik A. Roth. 2004. *Seeing What's Next.* Boston: Harvard Business School Press.

Covey, Stephen R. 1989. *The 7 Habits of Highly Effective People.* New York: Simon & Schuster.

Covey, Stephen R. 1991. *Principle Centered Leadership.* New York: Summit Books.

Fast, Julius. 2002. *Body Language.* New York: M. Evans.

Fleddermann, Charles B. 2004. *Engineering Ethics.* London: Pearson Prentice-Hall.

Flynn, Stephen. 2007. *The Edge of Disaster.* New York: Random House.

Gaynor, Gerard H. 2002. *Innovation by Design.* New York: Amacom.

Hirsch, Sandra and Jean Kummerow. 1993. *LifeTypes.* New York: Warner Books.

Hudson, Frederic M. 1999. *The Adult Years: Mastering the Art of Self-Renewal.* San Francisco: Jossey Bass.

Johnson, Robert. 1999. The spirit of leadership. *The Leadership Circle.* www.tlccommunity.com.

Kavetsky, Robert A., Michael L. Marshall and Davinder K. Anand. 2006. *From Science to Seapower.* College Park, MD: Calce EPSC Press.

Kawasaki, Guy. 2004. *The Art of the Start.* London: Portfolio.

Kelley, Robert E. 1998. *How to Be a Star at Work.* New York: Three Rivers Press.

Mahbubani, Kishore. 2005. *Beyond the Age of Innocence.* New York: Public Affairs.

Manz, Charles C., and Henry P. Sims, Jr. 1990. *Superleadership.* New York: Berkley.

Miller, John G. 2001. *QBQ: The Question behind the Question.* Denver: Denver Press.

National Academy of Engineering. 2004. *The Engineer of 2020.* Washington, DC: National Academies Press.

National Academy of Engineering. 2005. *Educating the Engineer of 2020.* Washington, DC: National Academies Press.

New Green Economy Council meets. *ENDS Report,* February 23, 2011.

Northouse, Peter G. 2004. *Leadership: Theory and Practice.* Thousand Oaks, CA: Sage.

Oakley, Ed and Doug Krug. 1991. *Enlightened Leadership: Getting to the Heart of Change.* New York: Simon & Schuster.

O'Hair, Dan, Gustav W. Friedrich and Lynda Dixon Shaver. 1995. *Strategic Communication in Business and the Professions.* Boston: Houghton Mifflin.

Owen, Hilarie. 2000. *In Search of Leaders.* New York: John Wiley.

Pastin, Mark. 1986. *The Hard Problems of Management: Gaining the Ethics Edge.* San Francisco: Jossey Bass.

Pinchott III, Gifford. 1986. *Intrapreneuring.* New York: Harper & Row.

Roberts, Wess. 1987. *Leadership Secrets of Attila the Hun.* New York: Warner Books.

Sheppard, Sheri D., Kelly Macatangay, Anne Colby and William M. Sullivan. 2009. *Educating Engineers.* San Francisco: Jossey Bass.

Surowiecki, James. 2004. *The Wisdom of Crowds.* New York. Doubleday.

3M earns ENERGY STAR award for sixth consecutive year. 2010. *Business Wire* (New York), March 18.

Utterback, James M. 1994. *Mastering the Dynamics of Innovation: How Companies Can Seize Opportunities in the Face of Technological Change.* Boston: Harvard Business School Press.

Wheatley, Margaret. 1999. *Leadership and the New Science.* San Francisco: Jossey Bass.